Super Natural

ALEX RILEY is an award-winning science writer who covers topics in conservation, evolution and human health. For PBS's NOVA Next, he trekked through the Puerto Rican jungle in search of a golden frog that hasn't been heard since 1987. For the BBC, he reported on a pair of beavers whose lives helped decide the fate of their kin throughout the UK. And for *New Scientist*, he visited Indonesian coral that were living and breeding in a simulated ecosystem in south London. He has a masters degree in Zoology from the University of Sheffield and held a research position at the Natural History Museum in London between 2014 and 2015. His first book, *A Cure for Darkness: The Story of Depression and How We Treat It* (Ebury, 2021), was a science-led investigation into his family history of mental illness and what treatments have ebbed and flowed through time and across the world. Alex's niche can be found in south Devon with his family.

Super Natural

How Life Thrives in Impossible Places

Alex Riley

Atlantic Books
London

First published in hardback in Great Britain in 2025 by
Atlantic Books, an imprint of Atlantic Books Ltd.

1 3 5 7 9 8 6 4 2

A CIP catalogue record for this book is available from the British Library.

Hardback ISBN: 978 1 80546 078 7
E-book ISBN: 978 1 80546 079 4

Text design and typesetting by Tetragon, London
Printed and bound by CPI Group (UK) Ltd, Croydon, CR0 4YY

Atlantic Books
An imprint of Atlantic Books Ltd
Ormond House
26–27 Boswell Street
London
WC1N 3JZ

www.atlantic-books.co.uk

MIX
Paper | Supporting responsible forestry
FSC® C171272

Product safety EU representative: Authorised Rep Compliance Ltd., Ground Floor, 71 Lower Baggot Street, Dublin, D02 P593, Ireland. www.arccompliance.com

To Nieve, William and Timothy

Living creatures press up against all barriers: they fill every possible niche all the world over... We see life persistent and intrusive – spreading everywhere, insinuating itself, adapting itself, resisting everything, defying everything, surviving everything.

SIR JOHN ARTHUR THOMSON, 1920

Contents

Introduction

The plastic dish is etched with a criss-cross of scratch marks, each catching the light like threads of spider silk on a damp and sunny morning. Snot-green clouds float by and then return, all moving together as if connected by an invisible force. One morning in September, I'm sitting in a laboratory at the University of Plymouth, looking at the contents of a Petri dish through a microscope. I turn the knob to adjust the focus, zoom in and out with the one that sets the magnification. Within a thin layer of spring water, there is a healthy population of tiny animals that I have come to see, each less than a quarter of a millimetre in length. They are also transparent. I rotate the dish and watch as the green clumps of algae jolt one way or the other at the slightest touch. Every movement is magnified. I make sure the scratches are in focus as these are what the animals cling to with their tiny, bear-like claws.

I can only see algae, the snot-green food. I feel a little bit like the young boy in *Jurassic Park* who peers through his fancy binoculars, hoping to see a T-rex, but only sees a goat chained to a pole.

After a few minutes of turning and scrolling, I start to wonder whether there are any animals in here at all.

Ellis Moloney, a PhD student here at the University of Plymouth, picks up my Petri dish between thumb and index finger and places it onto his microscope. 'There are loads in here,' he says excitedly, then quickly reassures me that his undergraduate students struggle to find them too. A child who grew up fascinated with the strange life of the deep sea, of transparent comb jellies and wide-mouthed gulper eels, Moloney has a habit of focusing on places other people might find lifeless or boring. When I met him, he's wearing a Pearl Jam T-shirt, light-brown cords and sporting a ginger-blond moustache. He is slight-framed but has an unexpectedly booming voice. He also swears a lot, both when discussing his laboratory animals – 'they f***ing die all the time' – to when he's awestruck by the magnificence of nature: 'those deep sea worms that are taking in all the f***ing sulphur from hydrothermal vents or whatever, and they're loving it'.

He hands the dish back to me and I soon find them, guided by his description and the fact that they are definitely there to see. A small worm-like blob, reminiscent of a caterpillar but quite unlike anything I've seen moving through soil or over a leaf, peacefully bumbling over a piece of brown sand. It rocks from side to side, pulses up and down, and makes me think of someone trying to swim in a pool of beach balls. This is *Hypsibius exemplaris*, a species of tardigrade. Also known as water bears, these microscopic animals have become famous for their ability to endure inhospitable conditions. Animations and videos of these tiny critters amass millions of views online. As one such video states, they have become the 'celebrity of the microcosmos'.[1] This is undoubtedly a result of their superhero-like abilities to withstand unimaginable extremes – freezing to near absolute zero, boiling heat, pulverizing radiation, the vacuum of space – that would kill a human in seconds to minutes. But there is another, equally important, reason for their fame: they are damn cute.

This quality has been recognized for nearly as long as we've known about tardigrades. Writing in his 1861 book *Marvels of Pond Life*, Henry James Slack made this important contribution to science: 'a little puppy-shaped animal very busy pawing about with eight imperfect legs, but not making much progress with all his efforts… a very comical amusing little fellow he was.' Looking through my microscope lens, I can't see in enough detail to fully appreciate the puppyishness of this particular tardigrade. But I know it's there. With rounded backsides, flattened faces, a way of moving that is so awkward as to look childish, water bears are more reminiscent of teddies than grizzlies.

Surprising, then, that such a squishy and microscopically cuddly animal would turn out to be so extraordinarily tough. Again, the supernatural powers of tardigrades have been known for a long time. As one tardigrade enthusiast, a priest living within a religious community at Techny, Illinois, noted in 1938, they have an 'extraordinary force of resistance', surviving half an hour in boiling water, seven months of freezing at −200 °C in liquid helium, pressures of 1,000 atmospheres (think Mariana Trench), intense radiation (UV, radium and X-rays), as well as a range of toxic gases.

Captivated, the priest concluded that it was 'impossible as yet to state what does eventually destroy their lives'. More recent studies have replicated this early work, demonstrating that water bears can endure temperatures that are just a smidge above absolute zero (−273.15 °C), the lowest temperature possible and a state in which molecular motion of any kind stops. At the opposite extreme, temperatures of 151 °C don't kill these creatures (although there are exceptions, with some species being quite sensitive to heat). Radiation levels a thousand times the lethal dose for a human can be shrugged off as if it were a bit of sunburn.

Their powers for survival are truly out of this world. In September 2007, tardigrades were sent into space inside a thick metal capsule known as Photon-M no. 3, a sphere of metal that opened as they entered low-earth orbit (250 to 290 kilometres above the ground). Exposed to the vacuum of space – its extremely low pressures, freezing temperatures and unfiltered UV radiation – they travelled around the world at 7.8 kilometres per second and encircled the globe 192 times. On 26 September, the capsule entered Earth's atmosphere as a ball of flame, all the animals on board protected by a thick, 12-kilogram heat shield. After a final parachute to the ground, it was collected by helicopters and transported back to the Netherlands where the European Space Agency is based. In a sterile environment, at the hands of a silicone-gloved scientist, it was found that the tardigrades were largely unaffected by the vacuum of space, a deadly mix of low pressure and anoxia (no oxygen). Only when they were exposed to the full spectrum of cosmic radiation (UV-A, UV-B and ultraviolet) did they show significant mortality. Even in the harshest treatment group, however, a few hardy members of *Milnesium tardigradum* survived. 'Our results... represent the first record of an animal surviving simultaneous exposure to space vacuum and solar/galactic radiation,' wrote Ingemar Jönsson, professor of theoretical and evolutionary ecology at Kristianstad University in Sweden, and his colleagues in 2008. How they were able to not die, they added, 'remains a mystery'.

The tardigrade's powers of survival have become so renowned and respected that, in 2017, physicists from the University of Oxford and Harvard University used these animals as the ultimate marker of the apocalypse. 'For complete sterilisation [of the planet],' they wrote, 'we must establish the necessary event to kill all such creatures.' In their paper published in the journal *Nature*'s *Scientific Reports*, the authors calculated that only an asteroid of similar size to the two largest asteroids known in our solar system – Vesta and Pallas, both a thousand times heavier than the asteroid that wiped out the non-bird dinosaurs – would have the potential to destroy these animals. More than this, the paper – entitled 'The Resilience of Life to Astrophysical Events' – concluded that once life begins on a planet, it is likely to endure. Tardigrades are just one extreme example of life's resilience. 'Even the complete loss of the atmosphere would not have an effect on species living at the ocean's floor,' they wrote. 'Impact of a large asteroid could lead to an impact winter, in which the surface of the planet receives less sunlight and temperatures drop. This would prove catastrophic for life dependent on sunlight, but around volcanic vents in the deep ocean life would be unaffected.' We will meet the inhabitants of these hydrothermal ecosystems in Chapter 7.

And what about the current catastrophe for which we are responsible? How might a tardigrade fare in a warmer, more unpredictable world? In 2021, one experiment simulated the projections of climate change and found that tardigrades were impervious to the hotter, drier conditions in which they lived. Even up to the 5.5°C 'worst scenario' of warming that the Intergovernmental Panel on Climate Change includes in its models for the year 2100, there were no effects on the diversity or abundance of tardigrades living in the outdoor facility in Duke Forest, North Carolina. Able to endure rapid changes in temperature and aridity, they seem to be one of the few climate change-proof animals.

Through their sheer indifference to environmental change, tardigrades are a lesson in an oft-forgotten fact about life on Earth: it is resilient. Surviving through five mass extinction events – one in which 96 per cent of oceanic life died – it takes more than a frozen world, poisonous oceans and an asteroid the size of Manhattan slamming into modern-day Mexico to end its reign on this planet. As an all-encompassing biological entity, life is very hard to kill.

Surviving day-to-day requires resilience. But endurance over the ages is only possible with ingenuity – finding new ways of living, whether it's adapting to novel food sources such as the radiation released from an unstable atom or inhabiting places that are so inhospitable that few other creatures can live there.

While tardigrades can endure unimaginable hardships and wait for conditions to improve, there are other characters in this book that depend upon extremes to thrive. In the ocean's depths, seven kilometres below the traffic of ships, there are ghostly fish that thrive in pressures that would crush our lungs and cause our blood vessels to burst. In sun-baked sand dunes around the world, heat-loving ants sprint from their nests only when the temperature has become lethal to every other animal. In the icy waters of the Southern Ocean, vast colonies of fish flourish in temperatures nearly two degrees below freezing. Inside the exclusion zone of Chernobyl, fungi feed on the reactor that exploded in 1986.

In our hubris (and anthropocentrism), we have dared to call a place lifeless, only to find some of the densest ecosystems open up before us. We have set limits on the temperatures of habitability, only to find these thresholds smashed. All animals breathe oxygen, we claimed, only for scientists to discover animals at the bottom of the Mediterranean Sea doing just fine without it. And when we think that all life depends on the sun (like we do), entire ecosystems are revealed that have as much use for sunlight as we have for hydrogen sulphide.

Water, oxygen, food, freezing, pressure, heat, darkness and radiation – this is our journey through extreme environments, places where fundamental requirements for life are either absent or in excess. These places on Earth act as portals into distant planets and moons that were once thought to be inhospitable but now excite astrobiologists looking for life beyond Earth. The adaptations to environmental stressors also provide insights into treating human disease and the preservation of cells and organs. But as we switch between different, sometimes opposing, stressors, the central theme of the book – resilience and ingenuity – guides us.

The concept of a 'niche', a place where an animal lives, feeds, reproduces, competes and dies, is central to the discussion of extreme environments. Why would any animal find itself in a place without oxygen, without food, bathed in intense radiation? It's because their survival isn't just a case of their environment (the 'abiotic' factors such as temperature, elevation, climate) but also the threat of predation and competition (the 'biotic' factors). When the concept of the niche was introduced to ecology, its central tenet was that no two species can inhabit the same niche. As the ecologist Joseph Grinnell wrote in his 1917 study of the Californian thrasher, a grey and unremarkable bird that scours dry soil for insects and hops onto branches to pick at berries, 'It is, of course, axiomatic that no two species regularly established in a single fauna have precisely the same niche relationships.' Put another way, every organism requires a unique place in the world. Survival and endurance depend upon being different.

This drive for a unique way of life has been the driving force behind evolution's continual move towards the extreme. From the first microbe to evolve in the depths of an ancient ocean, life has extended its reach into new frontiers, harnessed the power of the sun, moved onto land, flown into the skies and sunk into the deepest trench. There were always places where no predator could follow, no competitor could compete, at least for a time. Guided by changes in the Earth's atmosphere and the movement of the continents, life unfurled into every open space and every crevice; a once boring world was inoculated with wonder.

There's no scientific term for life's pull towards the extreme. Writing in his 1957 book *The Immense Journey*, the anthropologist and science writer Loren Eiseley tried to capture this tendency as an emotional tug, 'life's eternal dissatisfaction with what is'. This 'persistent habit of reaching out into new environments', Eiseley wrote, has allowed life to adapt to 'the most fantastic circumstances'. Evolution is, by necessity, adventurous. Or, as Dr Ian Malcolm put it in *Jurassic Park*, 'Life finds a way.'

Since I was a child, I have found solace in nature, spending as much time as I could on the old dial-up internet to learn about animals that lived on distant continents or in past geological eras. I was a typical boy growing up in the 1990s, obsessed with dinosaurs, big cats and the nature documentaries of David Attenborough. But, unlike for many of my peers, this interest never dwindled. It only deepened. I realize now that it has been my escape from a world that is often as confusing as it is unpredictable. In 2021, a time of COVID infections and extreme weather events, I found myself turning to those creatures that can endure the harshest environments, researching their endurance and resilience in the face of near-impossible odds. As our world came to a standstill, a period known colloquially as the Anthropause,

I spent my time reading up about those creatures that were completely oblivious to our world. Gelatinous fish in the deepest trenches, water bears that stumble through fronds of moss, silvery ants on the hottest sand dunes: their indifference was soothing, a bigger picture that provided context at a time when our lives were so condensed.

It wasn't an intentional behaviour; only looking back now does the pattern reveal itself. Seeking those organisms that could brave unimaginable environments helped guide me out of a seemingly inhospitable moment. Pondering a species that can survive the unimaginable, I could imagine a future that is still living. 'Those who contemplate the beauty of the earth find reserves of strength that will endure as long as life lasts,' Rachel Carson wrote in *Silent Spring*, a book focused on how we are sterilizing vast swathes of the planet with pesticides.

There is no denying that our planet is being devastated by our actions. Melting glaciers, dead zones in the oceans, unprecedented wildfires: many animals are being pushed beyond their limits and are threatened with extinction. Some scientists argue we are in the middle of the sixth mass extinction on Earth, one to rival that of the asteroid that crashed into Central America some 66 million years ago. The examples in this book can't balance the scales, nor should they be used as an argument to give up on those species that are threatened with extinction. It is a reminder that there is an in-built resilience to life on Earth – whatever challenges are faced, it will endure. 'Hope,' Rebecca Solnit writes, 'is an embrace of the unknown and the unknowable, an alternative to the certainty of both optimists and pessimists.' In writing this book, the pessimist within me – the voice that tells me that the natural world is already doomed so let's just give up – has been silenced, for now. 'Optimists think it will all be fine without our involvement,' Solnit continues, 'pessimists adopt the opposite position; both excuse themselves from acting.' To contemplate life's resilience is to nurture our connection with hope, with action.

CHAPTER ONE

DRY HARD

Water

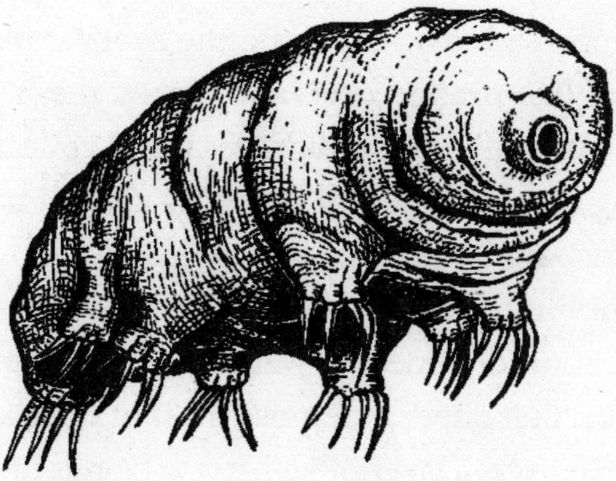

Let's begin with our poster child of life's resilience: the tardigrade. What makes them so invulnerable, so resilient, almost impossible to kill? How can they survive such a variety of lethal influences, from radiation to pressure, burning heat to freezing cold? The answer lies in their ability to live without that most precious resource of life, the sole search image for extraterrestrials used by NASA: water. Desiccation, it seems, is the greatest tonic against death.

The day after I had met Ellis Moloney in Plymouth, the rain still pouring down, I went looking for these animals in their natural environment. I was walking our dog Bernie along the coastal path near our home, a low-hanging grey sky closed in all around us. In sheltered coves, a barrage of white-tipped waves crashed on the rocks below. As we turned back, away from the coast, the sound of the sea was replaced by the roar of trees blowing in the wind. Soaking wet, I saw some thick pads of moss on an old dry stone wall, and decided this was the ideal place to find them. I scraped a few clumps with my fingernail, the rootless plants easily coming off like orange peel, and placed them in a clean poo bag. Although any tardigrades I had collected would be clinging to the moss with their tiny claws, I tied the bag with a loose knot to prevent any escapees.

Back home, changing out of my sodden shoes and clothes, I poured some tap water over the top and used some coffee filter paper to remove most of the excess water. What was left was a layer of soil, a few fronds of moss and, I hoped, a healthy population of water bears.

This wasn't my first foray in moss collecting. I had already bought a second-hand digital microscope – the kind that is often used when fixing broken electrical boards – to peer into the microcosmos surrounding my home. With a magnification of 40 times, it was perfect for spotting animals less than a millimetre long. But, alas, I had no luck finding water bears.

This time, however, my eyes had been trained to Moloney's lab-reared tardigrades. I knew the general shape, the way they moved, and focused on places that they could

PART ONE

Sustenance

cling to. Under the microscope, what looked like a thin layer of brown soil to the naked eye was transformed into discrete particles of sand. Brown was actually a range of hues from a pale blond to almost black. There were tiny fragments of moss and – there, yes there! – a small, transparent tube scrambling between the grains. At such low magnification, I couldn't see the finer details of this animal – it's pinprick eye, its eight chubby legs with claws at each end, the protrusible stylet that it uses to suck out the innards of algal cells. On my computer screen, I watched the blurry image and escaped the here and now for a few minutes. Observing an animal so indifferent to my existence was comforting. These minutes were a tiny portal into a world beyond humanity, and it was as simple as collecting some moss and buying an £8 microscope.

Simply by existing, these animals turn everyday flora – a piece of wet moss, a filigree of lichen – into a wonderland.

I took a blurry image of the tardigrade and sent it to Moloney. He replied the same evening telling me that it looked like a *Hypsibius* species, the same genus he works on in the lab.[1] Out of 1,380 species of known tardigrades,[2] the study of this diverse group of animals is largely limited to two members: *H. exemplaris* and *Ramazzottius varieornatus*. Similar in body form, recent studies have found that these two species represent two opposing methods of survival, especially in the face of radiation (as we shall see in Chapter 8). For now, it's enough to say that *Hypsibius* allows itself to be torn apart by extreme environments and then repairs the damage like *X-Men*'s Wolverine who can heal his wounds in seconds. *Ramazzottius*, meanwhile, protects itself from damage, using molecular shielding that allows it to endure radiation levels 1,000 times that which would kill a human. Repair and protect: a catchphrase for toothpaste but also relevant to tardigrades.

Able to survive extremes in heat, cold, radiation and pressure, most tardigrades are most comfortable on a frond of moss (hence their other, equally cute, name, 'moss piglets'). While this sounds like a plush place to call home, this micro-habitat is rocked by extreme fluctuations, sometimes daily. When dew from a cool night is evaporated by the heat of the midday sun, a waterworld becomes an arid desert. In order to survive, tardigrades would have to find a way of following the water, like a fish swimming from a seasonal stream into a permanent lake, or ride out the storm: accept change and embrace it. For the 500-plus million years of their existence, they have been doing the latter. Their name, tardigrade, means 'step slowly' for their clumsy and sluggish method of locomotion. Unable to swim, or move quickly in any way, they batten down the hatches when the water dries up. This daily extreme has led to some of the toughest animals on Earth, an ancient family of invertebrates that has conquered the world and tasted space.

Along with rotifers (also known as wheel-animals[3]), nematode worms and the larvae of a species of midge found in Central Africa, tardigrades enter a dormant phase in which they remove up to 98 per cent of their body's water.[4] Known as anhydrobiosis, or 'life without water', it is an exclusive ability that the word 'dehydration' doesn't seem fit to describe. Instead, scientists prefer to write that tardigrades become desiccated, a near-total loss of body water. Shrinking like a grape into a raisin, a squishy and transparent animal becomes a rigid husk, known as a tun, and in this state they are almost indestructible. Life without water, the tardigrade researcher Nadja Møbjerg wrote in 2021, 'seems to provide animals with a potential to survive conditions that are far beyond any constraints set by their normal environment'. For whatever reason, they are too tough. Surviving without the molecule of life – H_2O – makes every other stress seem like a breeze.

While anhydrobiosis is the term most often used, a slightly older term, and one of my favourite descriptions of this dried state, is 'chemical indifference'. Whatever happens outside its little bundle of desiccation, a tardigrade doesn't think or feel anything. It simply endures.

It also time travels. While in this dormant state, tardigrades don't seem to age. Months or years can pass by and they are still the same tardigrade they were when they entered their desiccated tun. In 2019, Lorena Rebecchi, Chiara Boschetti and Diane Nelson – three doyens of anhydrobiotic animals – called this an 'escape in time', a means of avoiding unfavourable environments when movement (an 'escape in space') is very limited.

When curled up into their little tun packages, tardigrades are so light that they are dispersed by the wind, like pollen. Able to move huge distances without even a functioning metabolism, if they land in a drop of dew they can reinflate and conquer a new territory. It's little surprise, then, to know that tardigrades are everywhere.

Tardigrades are one of the few animal groups – along with nematode worms, mites and a few other microscopic creatures – that can claim the distinction of living on every continent *including* Antarctica. On the Shackleton expedition of 1907–9, the biologist on board, James Murray, found that while the moss on this most southerly continent was 'dwarfed by the cold', there were masses of water bears in the freshwater lakes not far from the shore. As he wrote in his 'Tardigrade' monograph of 1910, 'the microscopic animals are not at all troubled by the rigours of the climate. When the cold comes they curl up and go to sleep, it may be for years, and when the thaw comes they go merrily on as though nothing had happened.'

In Greenland, another large landmass covered in ice, tardigrades live in pockets of liquid water – known as cryoconite holes – within glaciers. When sediment has been collected from the seabed, even in the deepest parts of the ocean, there have been tardigrades in it. On land, they are most often found on moss or lichen, but there are also species adapted to life among leaf litter and within the soil. Although the population densities change according to the richness of the humus, it isn't unusual to find 14,000 tardigrades in a square centimetre of soil. They have been found on mountain peaks reaching 6,000 metres above sea level.

Their diversity in habitats is reflected in their diversity in shape. Although tardigrades all share the same basic body plan of eight legs and one mouth, there are huge variations on this common theme. Many have claws but some species have evolved sucking discs on the ends of their legs. Most have flattened faces while others, such as *Bergtrollus dzimbowski*, a species first discovered in the Norwegian Alps, have long, anteater-like snouts that they use to slurp up their bacteria prey. But it is perhaps in the seabed where tardigrade evolution has had most fun. Living in the clean sands off the coast of the Faroe Islands, the brilliantly named *Tanarctus bubulubus* has 16 to 20 balloon-like organs attached to its rear end, structures that are used as buoyancy aids in the water and adhesive pads among the sand grains. Floating through the depths, they look like water bears that are always ready to celebrate a birthday party. Indeed, just knowing that *T. bubulubus* is floating somewhere in our oceans is worthy of celebration.

It is the species that live on land, in moss and lichen in particular, that are best known. Roughly 80 per cent of known tardigrade species are terrestrial, a figure that might be a result of their ease of collection, what's known as sampling bias, than actually a reflection of their true diversity. It is easier to collect a clump of moss from a tree than it is to sample the sediment from the deep sea. Whatever their number, the terrestrial species have long been the most interesting animals to scientists and amateurs alike, a microscopic insight into the remarkable resilience of biological stuff.

For much of the eighteenth century, water bears were at the centre of a debate of what life is. Since these animals were able to pause their vital functions for months and even years, a state with no perceptible metabolism, were they actually dead? Was it correct to say they survived without water? Or was the return of water not a re-emergence but a resurrection? 'The most relevant point,' one Italian scientist wrote in 1774, 'is to decide whether… the resurging little creatures in question are in fact truly dead, or only seem to be dead: assuming that they are truly dead, how does it happen that they come to life again?' In a time long before Netflix, these seemingly impossible reviviscences became a popular pastime: a drop of water on a glass slide could reveal a world of activity. Life without water, an animal impersonating death, became known as cryptobiosis – 'hidden life'.

How long could life hide before it was found – and consumed – by death? In 1753, Henry Baker found that nematode worms dried for four years could, with a drop of water, 'begin anew to actuate the same body'. He added that the limits for their survival were potentially limitless if their organs weren't 'broken or torn asunder'. '[M]ay they not possibly be restored to Life again… even after twenty, forty, an hundred, or any other Number of Years provided their Organs are preserved intire? This question future Experiments alone can answer.'

A few such experiments have now been conducted, although their conclusions are far from clear. In 1948, for example, Tina Franceschi dampened a 120-year-old sample of moss, finding that there were several tardigrades inside. After 12 days of observation, she noted a 'partly extended specimen' started to quiver in several body segments. 'In particular, in the front legs an extending movement followed by retraction was observed.' As she concluded, 'an activity of life appeared, even if very slight'. Although hardly convincing, this observation has been used as evidence that tardigrades can survive for over a century in their tun state. An article in *New Scientist* in 1998 turned the slight movement that Franceschi noted in her sample of moss into the statement, 'tardigrades were later found crawling all over it'. It can be forgiven to use hyperbole with these superlative-soaked animals.

More recently, samples from museums around the world have found that tardigrades may survive for up to ten years without water, still an extraordinary feat. As with the food we eat, the longevity of these dried animals can be increased if they are frozen. (In 2023, a nematode was unearthed from permafrost at Duvanny Yar, in the northeastern Arctic. It had been in a state of cryptobiosis for 46,000 years, a time when Neanderthals and woolly mammoths were still walking the Earth.) In 2016, two tardigrades were found in a sample of moss collected from eastern Antarctica in 1983. Frozen at −20°C for 30 years, they both re-emerged when rehydrated. Researchers from Japan's National Institute of Polar Research called them Sleeping Beauty 1 and Sleeping Beauty 2.

While Sleeping Beauty 2 died soon after its defrosting, Sleeping Beauty 1 started to eat the algae provided, filling her transparent body with seaweed-green blobs. Just over three weeks after rehydration, she started to lay eggs into her carapace, shedding this tough exoskeleton to act as a cocoon for her offspring. Over 45 days, she laid 19 eggs in five clutches, 14 of which went on to hatch, each imbued with their mother's uncanny potential to endure.

Once an arcane topic for the invertebrate taxonomist, water bears are now studied by scientists from around the world. Ellis Moloney at the University of Plymouth is just one member of a growing troupe of scientists hoping to reveal some of the biological mechanisms involved in tun formation, and the chemical indifference that it bestows on water bears. His supervisor Chiara Boschetti studies both rotifers and tardigrades to understand how biological material can be postponed in this dried state, a field that could lead to new means of keeping cell lines or vaccines viable for longer, and without freezing. 'It's just fundamentally fascinating,' says Bob Goldstein, a biologist who has studied tardigrades for over 20 years. 'Not only should they be dead but what they're made of should be destroyed too. DNA and proteins, which are way hardier than [cell] membranes or RNA' – the molecules that translate DNA code into recipes for proteins – 'would be destroyed by a lot of the conditions that tardigrades can survive.' It might be said that tardigrades, when in their protective tun, hold the fundamental ingredients of biology together.

Over 250 years since they were first discovered, scientists are still trying to figure out how tardigrades do this. In the 1980s, studies into other animals that could undergo anhydrobiosis, such as brine shrimps (also known as sea monkeys) and nematode worms,[5] revealed that a sugar called trehalose (two glucose molecules stitched together) is key to surviving desiccation. By swapping water molecules with this simple sugar, the cells of these animals could be kept structurally sound as the body shrinks and twists. In the absence of water, trehalose can turn into a biological form of glass, turning a delicate liquid into a tough solid. (Pumping cells full of sugar is a strategy for surviving being frozen solid for some frogs, as we shall see in Chapter 4.) More recently, with doubt arising over the importance of trehalose, so-called chaperone proteins have come to the fore. Late Embryogenesis Abundant (LEA) proteins, Secretory Abundant Heat Soluble (SAHS) proteins, Cytoplasmic Abundant Heat Soluble (CAHS) proteins: all have instantly forgettable names and, for this story, it's enough to know that they all do very similar things. Found in plant seeds as well as animals, they act like the obsessive-compulsive groundskeepers of a cell, holding everything in a particular place as the world around begins to bend out of shape. During desiccation, cell membranes and proteins tend to become misshapen or stick together. Since all reactions of life require movement and key-to-lock precision, these chaperone proteins support a cell through hard times.

Far in the future, Moloney hopes that his work can help with human disease. Parkinson's, Alzheimer's and most other neurodegenerative diseases, he tells me, are all associated with a cell's response to stress. If there are genes for resilience only found in tardigrades, could these be introduced into the dying cells of the brain? 'It's sort of like a pipe dream, but it doesn't have to be,' he says. 'There just needs to be more funding to accelerate the research. Because the potential, for sure, is there.' Intense radiation, extreme dehydration, freezing cold or burning heat: the slow-stepping water bear might hold evolution's Rosetta Stone for survival. As we shall see in Chapter 8, DNA from tardigrades has already been inserted into plant cells, bacteria and other cell lines, imbuing them with greater resilience in the face of environmental stress.

Anhydrobiosis is the most extreme adaptation to water stress, a forestalling of life when water vanishes. But coping with water scarcity, more generally, is found across the tree of life.

Some of the most spectacular examples come from plants; so-called xerophytes (literally, 'dry' and 'plant') live in scorching deserts and free-draining cracks in cliffs, and can spend months in a dormant state – similar to tardigrades – only to then re-emerge with the return of water. There are only 330 known species of such 'resurrection plants', a tiny proportion (0.086 per cent) of the 383,671 known plant species that grow stems and trunks and branches (vascularized plants). As time-lapse videos show, a plant that has shrunk into a brown tangle of twigs can reinflate, turn green and recommence photosynthesis within a few hours of being watered. It's not that they survive desiccation by re-emerging anew from a root system; rather, it looks like they are bringing their leaves back from the dead.

As with tardigrades, one of the tricks that plants utilize is to produce simple sugars that snap into a glass-like form when dehydrated. This solid phase reduces the chemical reactions that go on inside the plant's cells and also protects it from physical damage caused by shrinking. But this happens quite late in the process for plants, at the point where 90 per cent of the cell's water has already been lost. The first problem they encounter when water disappears is that plants don't just have cell membranes (like animals), they also have cell walls. Surrounding the cell membrane like the perimeter of a castle, this structure doesn't flex as the water disappears. It is too rigid – hence 'wall'. For resurrection plants, it has to fold, concertina-like, enabling the cell to reduce its size by over 80 per cent and suffer no damage. When water returns, the cell wall, still in contact with the more flexible interior cell membrane, unfolds into its previous state. Plants that aren't tolerant to water stress simply tear themselves to shreds, the internal membrane ripping itself free from the immovable wall that surrounds it. This is a recipe for plant death.

So is photosynthesis, at least when there's no water. If this light-harvesting process is allowed to continue in its absence, a plant's cells will start to churn out harmful reactive oxygen species (or ROS), a suite of metabolic by-products that can tear through the cell's innards, including its DNA. To stop this, the leaves of a resurrection plant fold into themselves, exposing their undersides to the sunlight and shading the more delicate upper surface. Packed with a natural form of sunscreen – known as anthocyanins – they reduce the amount of sunlight the plant's chloroplasts can absorb, ramping down photosynthesis as the plant continues to dry out. Some resurrection plants even digest their chloroplasts entirely, preventing photosynthesis from happening and reducing the damage from ROS to a minimum. When the water returns, they rebuild their photosynthetic machinery from the bits and pieces they started with.

Incredible to behold, the tricks used by resurrection plants are actually common to all plants. In recent years, work by Jill Farrant at the University of Cape Town and her colleagues have discovered that these desiccation-tolerant species are using the same mechanism that other plants use in their seeds. They simply lose this ability after they germinate. By genetically turning these genes back on in adult plants, Farrant and her colleagues wonder, can they flip a switch in their development, turning a desiccation-intolerant species into a tolerant one? Since 50 per cent of the world's plant-derived energy comes from three plants – wheat, rice and maize – can these major crops be reprogrammed to cope with periods of intense and prolonged drought? Since the amount of land that is too dry for agriculture is forecast to spread across the United States, Africa, Southern

Europe and Australia, can resurrection plants become the inspiration for what Farrant calls 'climate-smart agriculture'? While there will be lower yields due to periods of dormancy, such crops would be able to grow on land that would simply destroy the current strains of maize, rice and wheat. As Farrant wrote in 2022, 'This strategy, albeit extreme, guarantees the survival of [crops through] the harshest drought.' Can the future of agriculture be as simple as reminding a plant of its time as a seed?

A plant doesn't need to come back from the dead to live in the desert. Another xerophyte has long fascinated botanists for its ability to remain green in one of the driest places on Earth: the Namib Desert of Southwest Africa, one of the few 'hyper-arid' regions of the Earth where rainfall is less than ten centimetres per year. Along this thin stretch of coastline, the annual rainfall might be limited to a single shower during a thunderstorm. And yet, billowing up from the gravel and sandy plains that stretch between ridges of wind-polished marble, are thousands of giant plants that have been described as 'a stranded Octopus on a bare desert surface'. A tangle of tattered leaves emerging from the barren soil give the impression of a many-tentacled cephalopod. But the trained eye of a botanist soon realizes that *Welwitschia mirabilis* has only two leaves emerging from a central node, each one battered and torn by the harsh, dry winds of this coastal ecosystem. For this reason it has also been called a 'seedling in arrested development'; while other plants grow new leaves that unfurl from stems and branches, *Welwitschia* simply makes its first leaves bigger. Slowly, over centuries of its life, they can grow into tangled giants with each leaf extending to more than six metres (if they were straightened out) along the dry ground.

The size of the leaves adds to the oddity of *Welwitschia*. The textbook example of a desert plant is a water-filled chamber studded with leaves that have turned into tiny needles. A cactus. This is a great way to store rainfall and reduce water lost through evaporation. And while trees growing in arid grasslands, such as the baobab of Madagascar, send taproots deep into the Earth to slurp up groundwater, *Welwitschia*'s roots don't go much deeper than a metre. Gillian Cooper-Driver, a botanist writing in 1994, put it this way: 'It's been said that if botanists were to invent the ideal plant for a desert environment, surely they would never come up with a monster like *Welwitschia*.'

And yet, in the Namib, there are over 50,000 of these plants dotted across their eponymous ecosystem: 'the Welwitschia Plains'.

Welwitschia are sustained by rainfall, even in places where rain might fall just once a year. Their roots are shallow but dense, soaking up as much water as possible before it sinks or evaporates under the sun's heat. Their photosynthesis is incredibly efficient and the holes in their leaves – known as stomata – don't let much water escape into the dry and hot air. No taproots, no resurrection, just two green leaves becoming more untidy with time. Still partly a mystery as to how they survive, it's undoubtedly a resilient combination in the face of water stress. The Afrikaans name for this plant, *tweeblaarkanniedood*, means 'two leaves that cannot die'.

Away from the gravel plains, the sand dunes of the Namib might seem relatively lifeless. No plants anchor into the forever shifting sand. Vibrant green isn't a colour of this ecosystem's palette. But there is water here, if you know where and when to look for it. While rain may never fall in a 12-month period, the dunes are often blanketed by a thick fog that rolls in from a cold Atlantic Ocean current that has its origins in Antarctica. With this process occurring once a month in winter and five or six times a month in summer, a few species of black beetle, members of the Tenebrionid clan, have evolved small bumps across their backs that can accumulate the microscopic water droplets of fog into five-millimetre drops that can roll into their mouths. Without this bumpy surface, any water that collected on their backs would be lost to the heat and wind of the desert. With them, a beetle can consume a third of its body weight in water in one sitting. While other animals, from herbivores to lizards, will lap up beads of fog that hang onto leaves and grass, these five species of beetle seek out their own supply. As one review paper from 2020 puts it, 'The beetles are obsessive about their fog-basking.'

Fog-basking isn't unique to these beetles. The thorny devil, a species of lizard that was scornfully named *Moloch horridus* for its 'repulsive' horned and warty skin, can catch water from fog or intermittent rainfall. The scales of these lizards curl into a network of straw-like capillaries, each able to carry water droplets towards the animal's mouth. Even if this lizard came across a puddle, it wouldn't be able to drink from it directly; its face is so flattened, an adaptation to eat ants (and only ants), that it can no longer slurp up water. And so it drinks from its skin. If feeling parched, a thorny devil can find a patch of damp sand, stand in it and soon feel quenched.

While the Namib beetles and thorny devils use their bodies to harvest water, *Lepidochora kahani*, another black and very similar-looking Namib Desert beetle, bulldozes a trench that runs down the dune either in a straight line or a maze of bends and dead ends. Whatever its shape, it is oriented perpendicular to the oncoming fog, a design that maximizes the amount of humid air it comes into contact with. By protruding from the dune's surface, the ridges of these trenches impede the flow of the fog, forcing water to accumulate in the sand. How the beetles then 'drink' this water is unknown. It isn't, as one study concluded, the 'ingestion of fog-soaked sand'. However it is done, these beetles clearly modify their environment in order to harvest its water. For this, they can be compared to a species that isn't a lizard or a beetle – us.

In arid places where fog is a regular part of the weather system, huge mist nets have long been used to collect water for irrigating crops or as fresh drinking water. In Lima, Peru, for example, the fog locally known as La Garúa has been harvested to grow figs, grapevines and olives in places that were once defined by drought. Similar structures are found across the Pacific Coast of South America, the Middle East and the Namib. But while humans may have been harvesting fog for centuries, beetles and lizards have been doing it for millions of years. By studying the microstructure of their wing covers and scales it is hoped that even more efficient mist nets can be constructed, without a substantial increase in cost. A 2019 study found that mimicking the patches of water-attracting bumps and water-repelling troughs of the beetle's back could double the amount of water collection from fog, compared to a surface that is made of a plain water-repelling surface. As the water builds into substantially sized drops on the bumps (unlikely to be blown away in the wind) they then roll onto the hydrophobic surfaces along which they move quickly towards collection.

Out of 200 Tenebrionid beetles that live in the Namib Desert, only five collect water from the fog that flows over the sand dunes. For a place where water is so scarce – a heavy rainfall can be a 40-year occurrence – this low number is still a mystery. One possibility is that it is only a lifestyle suited to the most isolated and barren places, dunes where predators such as insectivorous lizards are unlikely to pop out for an evening patrol and find a platter of moist beetles with their heads in the sand. While it is undoubtedly an ingenious solution to water scarcity, it does make them incredibly vulnerable to being eaten.

How these beetles know when fog is about to roll in is also still a mystery. Could it be a change in air pressure? The humidity of the previous day? Whatever the case, when a row of beetles are face down on the sand, it looks an awful lot like they are praying to their water deity.

The driest regions of our world – from arid grasslands to sandy deserts – may seem like outliers and alien landscapes to us. In the hottest deserts and driest plains, we can't tend to cattle or grow crops. As a species, we live primarily in regions where water is available for at least part of the year. In terms of land area, however, arid regions account for up to 40 per cent of Earth's continental surface. In Antarctica and the Arctic slope of Alaska, water is frozen and not available in its liquid form. Nearer to the equator and in the shadows of mountains, hot air descending from high in the atmosphere prevents colder air from precipitating into rain. The Atacama Desert of Chile sits between two rain shadows, one from the Andes and the other from the Chilean Coast Range, a geological circumstance that has made this strip of coastal desert the driest place on Earth (excluding the polar deserts of Antarctica). Rainfall can be as little as three millimetres per year. As with the Namib, it is a hyper-arid landscape. The Atacama is so consistently dry that it is used as a benchmark to which other deserts are compared. Move in any direction from this thin stretch of South American coastline and it can only get wetter.

The Atacama has been referred to as 'a territory so bleak and desolate [that it is only] distinguished by the number of its hideously barren hills of rock and its sandy wastes'. Visit at the right time, however, and even this hyper-arid desert can be a riot of life. Over 1,000 species of plants have been found within its borders, many of which lie dormant as seeds for years and can sprout into seedlings at the slightest patter of rain. As one botanist has noted, 'As if aware that they have an ephemeral life and that what they have to do must be done quickly, they are scarcely above ground before they put forth blossoms.' Dense masses of yellow, cup-shaped flowers, cushions of the chamomile-like *Closia*, silk-like stems that tangle into what's known as Angel-hair, cacti topped with broad flowers the colour of lavender: even the harshest desert can adorn itself in delicate finery.

Seeds allow plants to lie dormant until the time is right for growth. They also allow a very shy mammal to live a full and healthy life without ever needing to stop for a drink.

Merriam's kangaroo rat of the southwestern United States is most notable for its extremely long legs and its characteristic hopping form of locomotion which it shares with its larger, Australian namesake. With a long, feather-like tail, it jumps from place to place and does not scurry like other rodents such as mice or rats. Able to fit in the palm of your hand, its jet black eyes are huge for its body size, an adaptation to its strict nocturnal lifestyle. Even a moonlit night is too much illumination for these seed-eating mammals, a diffuse spotlight that owls, foxes and coyotes use to their favour. If attacked, however, the kangaroo rat's long legs can springboard its lightweight body from potential harm. Even a photographer trying to capture an image of this animal is often met with a blank frame or a blur next to a hole in the ground. Between the click of the shutter and the flash, the animal has already leapt for cover, 'a disappearance into the burrow so sudden as to be almost startling', the 1922 paper 'Life History of the Kangaroo Rat' states.

But to focus on its legs as the essential adaptation to life in the patchy and bare desert is to ignore a hidden talent of Merriam's kangaroo rat. These rodents never drink water. Its daily life is defined by this absence. Not only is there no water to drink, it subsists almost entirely on dry seeds from creosote bushes. 'They do not feed on cactus, as some other rodents do, and they seldom eat green leaves,' one author notes. 'This is a very unusual performance for a mammal.' So how does it keep its cells, its blood, its brain hydrated? First, even a sun-dried seed still contains some water. Even if it is as little as 5 to 10 per cent of its weight, this can be a valuable drop of water in the desert. Kangaroo rats are so attuned to the need for water, they can select between two seeds that differ by 0.0014 millilitres, and then choose the ever so slightly wetter one. Second, there is another water source that the kangaroo rat can utilize. It doesn't come from its environment but from inside its body. Water is a by-product of metabolism, the breakdown of proteins, fats and carbohydrates. Let's take glucose, a simple sugar, as an example. Its chemical formula is six carbon atoms, 12 hydrogen atoms and six oxygen atoms. It can be written like this: $C_6H_{12}O_6$. When burned in the presence of oxygen inside an animal's cells, these carbons and hydrogens and oxygens are transformed into carbon dioxide (CO_2) and water (H_2O), plus energy. The whole process looks like this:

$$C_6H_{12}O_6 + 6O_2 \rightarrow 6CO_2 + \mathbf{6H_2O} + \text{energy}$$

(glucose + oxygen → carbon dioxide + water + energy)

Notice the water in bold? From just one molecule of glucose, it turns out, an oxygen-based metabolism can produce six molecules of water. For most animals, this so-called metabolic water has no impact on their water budget. In order to produce it, they need to breathe more oxygen, and the act of breathing means that water is lost through their lungs and noses. But kangaroo rats have such small and efficient noses that they cool their nasal passages with every inhalation, creating a surface for the humid air from the lungs to condense onto with every breath. It has been called a 'counter-current heat exchanger', a way for the kangaroo rat to hold onto the water it produces from its own metabolism. As with the Fremen in Frank Herbert's *Dune* novels, preserving water in the desert is a priority and a privilege. To lose or waste it is sacrilege. But the stillsuits that the Fremen wear to recycle the water in urine, sweat and breath have to be worn properly and maintained. For animals like the kangaroo rat, it is part of their being.

There are other ways to hold onto precious water. The long tubes of their kidneys concentrate their urine into a dark broth, reducing the amount of water lost as they excrete urea (a toxic by-product of digesting protein). In fact, their kidneys are so efficient at holding onto the water inside their bodies that Merriam's kangaroo rats can drink seawater and still keep their bodies hydrated. If we were to do the same, the salt in seawater would pull water out of our cells by osmosis and push it into our urine, slowly dehydrating us. Only Merriam's kangaroo rats living in captivity have been given such a strange option as seawater.

Hopping from patches of creosote bush, a kangaroo rat stands on her two hind legs while her short front paws pick up dried seeds and pack them into two fur-lined cheek pouches. Her large eyes and ears constantly sense the world for danger, but she lives entirely without fear of running out of water.

A kangaroo rat's ability to survive without drinking water is as much about behaviour as physiology. To avoid the stiflingly dry air of the desert, they spend the day underground, cocooned in a much cooler and more humid environment. If they were forced up onto the surface, a place where humidity can be as low as 5 per cent, they would die in minutes. The amount of water lost to the surrounding air – through sweat and breath, primarily – would lead to a process known as an 'explosive heat rise'.[6] As the blood plasma loses its water, it becomes thicker, harder to pump around the body by the heart. As the body struggles to shift warm blood from the body's core to its surface, flushing the capillaries of the skin with warmth, the animal has no means of shedding excess heat. The cycle continues and deepens: more heat, less water, thicker blood, more heat, less water, and on and on. Even a kangaroo rat can't endure the daytime desert.

But a camel can. These large domesticated ruminants can lose 30 per cent of their body's water and still show very little discomfort (just 12 per cent would kill a human in the heat of the desert). This is because camels lose water differently, syphoning off the water from spaces in between cells rather than from plasma, allowing the blood to continue to flow around the body. They also allow their bodies to heat up to 41°C, a temperature that we would consider a potentially lethal fever. By letting go of their thermal regulation (and not trying to keep their body within a 2°C range as we do), camels lose less water in sweat, the main method of cooling down for many mammals. This is known as heterothermy (as opposed to our homeothermy), and is a common adaptation in large mammals that inhabit desert regions, saving four to five litres of water per day.[7] In the Arabian Desert, for example, oryx – large relatives of cows with metre-long horns – allow their body temperature to go from 36 °C in the morning to over 41 °C by the evening. And they can do this every day when it is hot and dry. At night, when the desert air once again cools down, these large mammals radiate this heat back into their surroundings by convection (and not evaporation, which is how sweat cools the body).

As with the kangaroo rat, a 100 kilogram Arabian oryx can subsist entirely on the water it consumes as part of its diet, largely the sun-dried leaves of *Disperma* shrubs. It prefers to forage during the cooler hours of the evening, when each leaf might be 30 per cent water. In the day, however, 'these plants are so dry their leaves fall apart when they are touched', Richard Taylor, a physiologist working in East Africa wrote in *Scientific American*. In the same essay, he noted how these animals are aggressive and willing to use their 'rapier-like horns with great facility'. Anyone who studies these animals, he added, will 'get physical as well as mental exercise'. For an animal that eats crumbling leaves and experiences body temperatures above 40 °C on a regular basis, it seems fair to be a bit cranky. Plus, in a time before digital biologgers, Taylor and his colleagues took body temperatures rectally.

Although there is a huge amount of variation and nuance in climate models, it is generally thought that dry regions of the planet will become drier while wet places will become wetter. Forever keen to find an acronym to fit their theories, scientists have called this the DIDWIW trend paradigm: 'drier in dry, wetter in wet'. As part of this trend, droughts are becoming more extreme, reducing vast areas of scrub or grassland to bare soil and rock. Even in megadroughts, however, there are animals and plants ready to benefit. In 2018, a study by Laura Prugh from the University of Washington and her colleagues studied the worst drought to hit the Carrizo Plain in California for over 1,200 years. From 2012 to 2014, rainfall in California was low but not exceptionally so. It was the intense, unrelenting heat that made water evaporate, turning the once verdant plain of grasses and shrubs into a 'barren plain nearly devoid of vegetation'. Since 2007, Prugh and her colleagues had been monitoring the abundance of 336 species of plants and animals, and how drought affected this ecosystem. Their datasets found that 85 of the 336 species (25 per cent) were classified as losers, experiencing a significant decrease in number with the lack of available water. But the vast majority – 71 per cent – showed no significant response to the drought, a core stability during a time of extreme change.

Surprisingly, 12 out of the 336 (4 per cent) were classified as winners, species that increased in number even after four years of intense drought. These included native red maids (*Calandrinia menziesii*), an annual species of plant that blooms in magenta-coloured flowers in places where invasive grasses don't dominate. Scurrying around these nutritious plants were increased numbers of common side-blotched lizards, six species of beetle, *Cyphamyrmex* ants, two birds (killdeer and the greater roadrunner) and, notably, the short-nosed kangaroo rat. Like its cousin Merriam's kangaroo rat, this small rodent hops on two large hind legs and eats seeds. Its success during the 2012–15 drought was due to a decrease in competition, a population crash in the dominant nocturnal rodent: the giant kangaroo rat.[8] After a year of drought, this latter species was unable to sustain its greater body mass, declining 11-fold over the period of drought. Drought, in this case, was a 'disturbance agent that opens niche space by stressing dominant species and allowing competitively inferior species to increase in abundance', Prugh and her colleagues wrote. In terms of rodents, species diversity actually increased during the drought, an ecosystem of many smaller-bodied animals rather than the dominant giant.

Whether this is good news or bad news depends on the way we envision the norms of nature. Do we treasure the larger animals that dwell in the wetter seasons? Or do the weedy plants and fast-growing, smaller animals of disturbance deserve as much of our appreciation? With droughts becoming more intense, especially in southern South America, Southern Europe and southwestern United States, it is these drought-adapted species that are likely to be the ultimate winners in a hotter – and therefore drier – world.

It is a world that annual killifish have evolved to thrive in. A hardy assortment of 320 or so species, these guppy-sized, gaudy fish inhabit drought-prone areas in South America and Africa and can complete their entire life cycle – from egg to egg-producing female – within 12 months (hence *annual* killifish). For their sudden appearance following rain, Indigenous communities have long referred to them as 'cloud fish'. In truth, their eggs were buried in the earth the entire time, wrapped in a protective cocoon that has allowed these fish to lie dormant, sometimes for two years, until the next rain. Dwelling in ephemeral ponds, annual killifish are marginal habitat specialists, able to survive droughts that would kill nearly every other type of fish. 'You can almost think of [these eggs] like a seed bank,' says Jason Podrabsky, a killifish researcher who has studied *Austrofundulus limnaeus*, a species of killifish found in coastal deserts and the savanna of Venezuela, since the 1990s. Except this is a vertebrate, a fish, and not a seed-producing plant.

Unlike the dehydrated tun of tardigrades, killifish eggs hold onto the water that they contain, shielding it from evaporation. Compared to similar-sized fish, annual killifish eggs have an 'exceptionally thick' chorion, a viscous envelope that makes the journey of water from inside the egg to the outside world much longer, and therefore harder. Plus, the layer surrounding this, known as the perivitelline layer, can lose all its water and turn into a form of biological glass, vitrifying a liquid bubble into a solid capsule. 'The embryos actually become like little marbles,' Podrabsky says. 'You can even drop them and they clink and they roll around. But if you look at them under the microscope, you can see the embryo inside twitching, you can see its heart beating every now and then.'

But every marble has its breaking point. For an annual killifish, water loss is inevitable. In the lab, Podrabsky has found that 50 per cent relative humidity led to killifish eggs drying up in a couple of days. At 75 per cent relative humidity, however, 40 per cent of the embryos could survive for over 100 days. While a relative humidity of 75 per cent sounds like a muggy day to us, for a fish that lives in water this is a dry breeze. In comparison, other species of fish that have evolved to deal with periods of drought require humidity levels close to 98 per cent to survive. In what has been called a 'summer sleep', the African lungfish buries itself into the lake bed as the soil above dries and hardens. Connected to the surface through a small breathing tube, they secrete a thick mucus layer, known as a cocoon, that reduces water loss. But without the damp soil surrounding it, the fish would shrivel and die. If it remains moist, they can survive in this state of 'estivation' for several years.

Annual killifish have made their most fragile life stage – the embryo – its most hardy. Being able to lie dormant through periods of intense drought has allowed these fish to inhabit marginal habitats that would otherwise be off limits for fish. (Walking through pastureland in Venezuela, Podrabsky has even found them in water-filled hoof prints made by cattle.) Once the rain returns, the embryos hatch, the adult bulks up on worms and the larvae of insects, reproduces daily and dies, all within two months. In the parched savanna of Southern Africa, the turquoise killifish, *Nothobranchius furzeri*, has been found to reach sexual maturity within 14 days, the fastest development of any animal with a backbone (a group that includes all jawless fish to amphibians, reptiles to birds and mammals). As the authors of that study wrote in 2018, 'Even a pool that desiccated in three weeks permitted successful reproduction of *N. furzeri* and, thereby, supported a viable population that year.'

The aphorism 'Live fast, die young' never found a more fitting mascot than annual killifish. Even if they live for two or three months, *N. furzeri* shows the classic hallmarks of ageing, from cognitive decline to cancer. Since other vertebrate models of ageing such as the mouse and the zebrafish can live to five years or more, this hardy species has the potential to fast-track our understanding of the biological mechanisms of ageing and death.

Depending on where you choose to begin the story of a killifish's life (the egg or the adult), it could be argued that the first inhospitable environment they have to survive is within the dark sediment into which they are laid by their soon-to-be-dead mother. Buried in a stagnant pool of water, microbial activity is high and oxygen content is low, perhaps even absent altogether. A 10-day-old killifish embryo can last for over a month without a whiff of oxygen. A month! After this period, one study concluded, there were no signs of developmental abnormality and the embryos followed a normal trajectory of growth.

And killifish aren't the only animals to survive for long periods without breathing oxygen. There are even fully grown adult animals, unprotected by a glassy marble, that can survive without breathing for up to six months every year. One group of animals might never breathe oxygen in their entire life cycle, from egg to adult, a phenomenon that, until 2010, was thought to be a defining feature of Kingdom Animalia.

CHAPTER TWO

BREATHTAKING

Oxygen

The small turtle kicks herself to the surface of the pond and pokes her tube-shaped nostrils into the frigid air of a November evening. A few bubbles emerge from her mouth before she inhales, filling her lungs with a fresh breath. The sun is setting and thin wisps of cloud are painted with orange and gold, colours not too dissimilar from the fiery stripes that adorn her keratinous skin. Soon, these sunsets will be hidden by an impenetrable ceiling. Every night, the temperature drops below freezing, and the hole in the ice she just breathed through is closing. By the morning, her pond will have frozen over completely and a forecast of heavy snow will add a reflective blanket that blocks any sunlight from penetrating into her home. The sun-dappled, algae-filled pond that was once an amphitheatre of croaking frogs and buzzing insects calms to a still silence. As the turtle sinks to the bottom, she settles like a flat pebble onto the soft sediment, a small cloud of silt blurring her vision for a moment. Then she waits. For sunlight, for food, and for her next breath. She might not fill her lungs again for another six months.

For most animals, oxygen – like water – is synonymous with life. Painted turtles that live in southern Canada, the northern limits of the species' range, seem to break with this tradition. It seems so extraordinary that scientists have done everything to try to prove that it mustn't be the case, that they must be getting oxygen somehow and from somewhere. Freshwater turtles can breathe through their skin and a specialized anal sac, for example. Could these overwintering turtles be absorbing oxygen over these dark winters? Is a lungful of air an anthropomorphism, a trait that we hold so close to breathing that we think it vital for all animals? This is certainly part of the story. If ice alone covers their ponds, light can still enter and power photosynthesis, a process that absorbs carbon dioxide and releases oxygen. Painted turtles can breathe through their skin and butts in such conditions, albeit at a reduced rate than if they were using their lungs. But when the snow falls the pond is pitch black. Any oxygen that remains in the water is taken up by the animals and bacteria (the decomposers) that feed on the remaining plant life. Hypoxia (a low concentration of oxygen) soon becomes anoxia (a complete lack of oxygen). In such conditions, there is no oxygen to absorb. A turtle is truly breathless.

In experiments in the late 1980s, one of the pioneers of freshwater turtle biology, Donald Jackson, found that even when he flushed his aquariums with pure nitrogen, thereby removing any oxygen contained in the water, the turtles were still very much alive. Three months passed with little problem. The turtles seemed healthy and unfazed. Two out of the ten turtles survived for six months in this state. 'The condition of these anoxic animals was quite poor, however, at this extreme stage,' Jackson added. The fact that they were alive at all was remarkable. And no one – not even Jackson – knew how they were able to hold their breath for so long.

An air-breathing animal that doesn't breathe for up to half a year sounds like an oxymoron, an adaptation that seems to go against the current of over 500 million years of evolution. Since animals evolved, their metabolisms have been powered by the combustion of oxygen, a gas that helped turn a microbial world into a multicellular wonder. When it comes to metabolism, there's nothing quite like it. For every glucose molecule that we digest from our food and shuttle into our cells, an oxygen-based metabolism can generate ten times more power compared to not using oxygen. This aerobic metabolism is like warming a room with a wood-burning stove. Anaerobic (i.e., without oxygen) is trying to do the same with a candle.

For much of life's four-billion-year history on Earth, anaerobic processes were fit for the task of maintaining and reproducing a single cell. But to explode into various arthropods in the Cambrian, dinosaurs in the Mesozoic or mammals in the Cenozoic, evolution needed a bit more of a kick. Oxygen is the fuel of complex life. To survive without it – for minutes or for months – is to push evolution out of its comfort zone. Ultimately, it shows the ingenuity and flexibility that is baked into evolution: an air-breathing animal can, for a time, live without the very fuel its cells require. In 2016, a study tested whether painted turtles could be trained to better survive prolonged periods of anoxia. As with exercising for a marathon, could repeated exposures to anoxia prepare their body for the coming stress? For two hours every other day, Daniel Warren submerged a few of his laboratory painted turtles to anoxia, stopping the experiment after 19 days. In mammals and birds, such repeated exposures can prepare the body for environmental insults such as bouts of freezing, extreme heat or oxygen stress. But his turtles showed no difference to turtles that had no prior exposure. Their metabolism was as efficient, their blood was suffused with the same amount of oxygen, and they recovered from their period of anoxia at the same rate. His conclusion? '[T]he physiological systems involved in tolerating and recovering from anoxic submergence stress in painted turtles are not plastic in response to repeated anoxic stress, but are, instead, constitutively adapted to survive extended periods of anoxia.'

To be a painted turtle is to be born with the ability to live without oxygen.

Turtles evolved some 230 million years ago in the Triassic Period, their earliest ancestors extending their ribs, generation after generation, until they formed their characteristic carapace. Their ability to endure periods without oxygen, however, is much more recent. To overwinter requires extreme seasonality: a fertile summer followed by a frozen winter. Only since the chill of the Pleistocene some 2.5 million years ago[1] hardened the northern hemisphere with seasonal ice cover did these animals begin to endure winters that would kill almost every other animal. From the frigid coast of Nova Scotia to the temperate rainforests of British Columbia, it is a prehistoric story of ice and fire-painted turtles.

If there's a definitive moment when science realized that there was something special about freshwater turtles living in the northerly latitudes, it was 1963, the year a paper by Daniel Belkin, a physiologist working in Florida, was published in the journal *Science*. The title was straight to the point: 'Anoxia: Tolerance in Reptiles'. Belkin tested 70 species from a range of reptile groups – skinks, iguanas, boa constrictors, vipers, geckos, a crocodile – and timed the period between their first and last breaths (they were resuscitated). All species tested were able to tolerate half an hour to an hour without oxygen. Compared to the few minutes a laboratory mouse could handle, these were impressive figures. But the turtles, Belkin found, were in a league of their own. The lowest score, measured in marine turtles, was two hours. The highest, an unnamed species of turtle belonging to the group that includes painted turtles, was 33 hours. As Belkin wrote, 'turtles are several times more tolerant of these conditions than are other reptiles'.

Why? Belkin didn't know, and neither did his peers. Later studies by Jackson in the 1960s, then a graduate student at Duke University, would find that they reduce their metabolisms by over 90 per cent during these periods of anoxia, lowering their body's demands when supply was cut off. But that wasn't unique to turtles. Most reptiles can do this. Even hibernating mammals can, albeit not to the same extent. With their metabolisms reduced to embers, it was clear that the turtles were using anaerobic metabolism, the process that we use when our muscles are pushed beyond their comfort zone. With their livers packed with glucose,[2] they could keep their cells ticking over during these times of anoxia. But how did they deal with the consequences? Anaerobic metabolism releases lactic acid as a by-product,[3] the feeling of numbness in our muscles. How could an animal cope with this build-up – an acidity that would poison their bodies, scrambling their neurons and crippling their muscles – month after month after month?

The secret was staring scientists like Belkin in the face. What do you see when you look at a turtle? Its shell. This armour plating of bone, made up of extensions of the ribs and the ossification of the skin, not only protects them from predators but also from the internal consequences of life without oxygen. In the late 1990s, Jackson discovered that turtles dissolve their bony shells into its constituent parts: calcium, magnesium and carbonate. The latter acts as an antacid, buffering the acidity that would otherwise build up from their slow, but ever-present, metabolism. 'We think of the [turtle's] shell as defending against an outer enemy,' Johannes Overgaard, a physiologist at Aarhus University in Denmark, says. 'But it's also defending against an inner enemy.'

Watching a group of painted turtles kept in laboratories at the University of Toronto, noticing the beautiful flashes of red and yellow that tip each scute of their shell, I knew that I wouldn't look at turtles in the same way again. Beautiful on the eye, a wonder for the mind.

I was visiting the lab of Les Buck to learn more about the research into these animals' tolerance to anoxia. Their ability to slow their metabolisms down to embers had been known since the 1960s, their shell-buffering system since the early 2000s. Buck began his research into anoxia in the middle of these two discoveries, in the 1980s, and he was primarily interested in how the turtle brain deals with such long periods of anoxia. Our own brain only weighs 2 per cent of our body weight but it consumes over 20 per cent of the oxygen we breathe. This is why a stroke, a blockage in an artery delivering oxygenated blood to our neurons, is so devastating. Our brains function on a rich supply of oxygen (and glucose). Take that away and it's like a fighter jet running out of fuel mid-flight. By analysing how individual neurons react to anoxia in the lab, Buck's research is starting to map the mechanisms that keep these cells alive for long periods without oxygen. Both when I met him and during our initial video call, he spoke a lot about GABA, the most common inhibitory neurotransmitter in the brain. Not a household name like serotonin or dopamine, GABA is the main 'stop' signal that most of our neurons use, shutting down the most common excitatory pathway in the brain: glutamate. Over 90 per cent of neurons that activate their neighbours into action – sending a signal of thought, memory or movement – use glutamate. 'We know GABA release is important,' Buck says. 'And [in turtles] there's a lot of GABA released.'

This took me a while to understand. It sounds counter-intuitive to actively control inhibition. Why not just turn everything off? Then I realized that we have a name for this: death. Life, fundamentally, is the control of imbalance, whether it's the flow of charge in our neurons or the calcium that floods into a muscle cell to initiate contraction. You can't just turn a brain off, you have to constantly keep it running on standby. This is what the turtles seem to be doing, for months.

Could these tricks of turtles be used for our own health? Could injecting GABA, or some other inhibitory chemical, reduce the damage caused by a stroke? What about a heart attack? Both are caused by blockages in an artery and lead to neurons or heart muscle being deprived of the gas they need to function. Without it, supply doesn't meet demand. Anaerobic metabolism isn't sufficient. A cell, sensing the fall in oxygen, panics and tries to increase its energy yield. In doing so, it begins to self-destruct. This is bad.

But it gets worse. Counterintuitively, the greatest amount of damage from a stroke or heart attack comes from the *return* of oxygen. Whether the blocked blood vessel is cleared naturally or a surgeon inserts a stent, reoxygenation isn't like a breath of fresh air but an ignition switch. It causes what is known as a 'reperfusion injury'. The details involve calcium and its sudden injection into the mitochondria, but there's one takeaway that summarizes it all quite neatly. 'It's a bit of a mouthful,' Gina Galli, a cardiovascular researcher who studies freshwater turtles, tells me, 'but it activates what's known as the "mitochondrial permeability transition pore". It's like the pore of death, basically. And that triggers everything to be killed.' Even the surrounding cells, unaffected by the blockage, can be destroyed in the aftermath. This is why Galli studies turtles. 'Why is it that they can be starved of oxygen for six months and reoxygenate and not have any problem at all?'

Scientific curiosity can soon shift into medical opportunity. 'After a while, you kind of start to think, "Well, is this something that we can really use to benefit society?"' Galli says. 'Are there pathways [in the turtle] that we can activate or deactivate in humans to try to do something about heart disease?' We're still at very early stages, she adds, almost at the beginning when it comes to understanding, never mind applying, the extraordinary abilities of painted turtles.

In Buck's lab, surrounded by the constant hum of filtration systems and the relaxing trickle of water, I peer into the clear waters and watch a few turtles grazing along the bottom of their jacuzzi-sized tanks. Others are laid on their bellies, the part of the shell known as the plastron, like sunken stones and show the same geological-level intent to move. And then there's the turtle on the rock in front of me, basking in the warm rays of a red UV bulb, watching me suspiciously with one eye. Like aquarium-kept fish, these turtles tend to follow any human that enters their room. They learn where the food comes from. But this particular individual doesn't seem so enthused at my presence. It tucks its head into its shell, the circle of wrinkly skin folding up like a leathery accordion. Its front legs follow, leaving only its back legs and pointy tail sticking out. The embodiment of indecision, I think. A turtle caught in two minds. Then, after a few seconds, it seems to decide that I was no threat, and closes its heavy-lidded eyes and, I assume, falls asleep. I was struck by the flexibility of this animal, still in that half-in, half-out pose. It is rigid but also retractable. Amphibious, able to live in water as well as on land. It breathes air but can survive for almost half a year – half its life – without it.

Extreme breath-holding has allowed turtles to extend their distribution into northerly climes that other reptiles can only dream of. It has also allowed mammals – a group of animals that evolved from reptilian ancestors on land – to return to the oceans. Hunting their giant squid prey, sperm whales can dive to 1,200 metres below the ocean's surface, holding their breath for upwards of an hour. The elephant seal, an animal famous for its bulk and beanbag-like movements on land, can hold its breath for two hours. The champions of mammalian breath-holding, however, are the beaked whales, a family of 21 species of cetaceans that have toothy snouts and are so secretive that some species have never been seen alive. The spade-toothed whale, for example, is only known from skeletal remains and a few stranded individuals that have been found on the shores of New Zealand in 2010, 2017 and 2024. As one paper puts it (with a whale-sized understatement), they are 'some of the world's most cryptic and difficult to study animals'. When they have been tagged with a data logger, they introduce scientists to new depths. In 2020, Nicola Quick, a marine biologist at Duke University, and her colleagues recorded a maximum dive time of three and a half hours. While this may have been a stress response to sonar, the scientists noted, 'These extreme dive durations… are perhaps more indicative of the true limits of the diving behaviour of this species.' (Unlike human free divers who hold air in their lungs when they dive, beaked whales exhale, emptying their lungs because they are inevitably going to collapse. Every time they surface, they reinflate. As someone who has punctured a lung and felt it collapse, I find this incredibly painful to imagine. Whales do this without harm many times a day.)

Although shorter in duration, minutes rather than months, each of these dives are no less remarkable than the overwintering of a painted turtle. These animals aren't lying in a semi-comatose state waiting for conditions to improve, they are active. And they're not just actively swimming, they're hunting, using their muscles and sensory systems to locate, catch and consume squid and fish. And, for the deepest dives, they are doing this in complete darkness. While beaked whales emit ultrasonic clicks and listen to their echoes to orient themselves, diving seals use visual cues such as the bioluminescence of their prey and, for the final attack, the super-sensitive whiskers – or vibrissae – on their faces. A camera attached to the head of an elephant seal captured the moment it pursued a *Taningia danae*, a two-metre-long squid with lemon-sized light organs at the end of two of its arms, a reminder that our perception of these animals as heavy beach lubbers couldn't be more insulting.

The same is true of penguins, the clumsiest of birds on land but the finest divers in the oceans. The emperor penguin (just one of 18 species of penguins that inhabit the southern hemisphere) is able to reach depths of 500 metres and hold its breath for half an hour.

Distant branches on the tree of life, all these animals have similar approaches to holding their breath while swimming. Inside their bodies are vast reserves of oxygen that can be used like the compressed tanks of a scuba diver. Instead of needing to breathe through a mask, however, oxygen is shuttled straight from their blood (from haemoglobin) and muscles (held by myoglobin) into the cells that need it, whether it be the fast-firing neurons in the brain, the constantly active optic nerve, or the fast-twitch muscles of fin, flipper or wing. The amount of myoglobin in their muscles makes their tissue look almost black. There's so much oxygen in storage that even the deepest diving whales and seals rarely have to swim past their aerobic limit, the moment their metabolisms switch to anaerobic pathways.

Elephant seals even sleep while holding their breath. In 2023, a study published in *Science* found that these animals swam to depths of 200 metres, a place where little light can penetrate, and then started to power down. 'Then they transition to REM sleep, where they flip upside down and spin in a circle, falling like a leaf,' Jessica Kendall-Bar, the lead author of the study, said at the time. Sinking in a downward spiral, belly facing upwards, the seals then snapped out of their power nap after ten minutes or so, swimming back up to the surface to breathe.

Mammals and reptiles can stop breathing oxygen for a period. But they are still connected to the gas through the myoglobin they store in their muscles or the oxygen they will need to breathe once spring melts the ice above their heads. Since its discovery in the eighteenth century, oxygen has been thought to be a vital ingredient to animal life, at least at some point in a species' life cycle. Then, in 2010, a simple but profound little critter shook this dogma.

In 2008, Roberto Danovaro and his colleagues from Marche Polytechnic University in Ancona, Italy, analysed sediment samples from some of the deepest parts of the Mediterranean Sea, a place over 3,000 metres below the surface that is eternally dark. Hidden among the gloom are mirage-like formations known as deep hypersaline anoxic basins (DHABs). As seawater mixes with an ancient deposit of salt under the seabed, it becomes saturated and heavy, a blanket of brine that sinks into the basins just as lakes fill depressions in a landscape. There is so much salt in DHABs that there is actually very little water. As one study puts it, 'DHABs are some of the driest places on Earth despite being located in the ocean.' Oxygen levels plummet from the upper layer of the brine and quickly become anoxic. Completely dark, dry and without oxygen, DHABs are often called one of the most extreme habitats found on Earth. And yet, Danovaro and his colleagues found not just extremophile microbes – bacteria, archaea, fungi – but tiny animals known as Loriciferans, another sandgrain-hugging member of meiofauna, the classification of microscopic animals that includes tardigrades and nematode worms. Unlike tardigrades, however, Loriciferans are only found in the ocean. And they are nowhere near as cute. Under the microscope, they look like spiky squids that have been turned upside down, hundreds of spindly tentacles surrounding their mouth.[4] First discovered and named in 1983 from specimens collected off the coast of France, there are now over 30 known species of Loriciferans. None are longer than a millimetre.

One of the species found in the anoxic sediment of L'Atalante basin, a DHAB nearly 200 kilometres directly west of the island of Crete, was new to science and, in 2014, Danovaro and his colleagues named it *Spinoloricus cinziae* after his wife, the marine biologist Cinzia Corinaldesi. Dive 3,000 metres into the Mediterranean, plunge into a lake of asphyxiating salt, and you can still find romance.

While previous reports from the Black Sea also found evidence of animal life in anoxic waters, these creatures were thought to have died in waters above and sunk into the depths, a process that was delightfully called a 'rain of cadavers'. But *S. cinziae* were *only* found in the DHABs, not in the oxygenated sediments surrounding it, suggesting that they didn't simply fall from the shallower waters but were actively – and exclusively – living in this inhospitable habitat. As Danovaro and his colleagues wrote in the journal *BMC Biology*, 'sediments in the neighbouring [oxygenated seabed] of the L'Atalante basin were also investigated at the time of sampling as well as in several other occasions since 1989, and we never found one single individual of the phylum *Loricifera*'. Moreover, since they found specimens that were moulting as well as ovaries holding eggs inside their bodies, they concluded that these animals were growing and reproducing in these waters, two life stages once thought to be dependent on oxygen and the energy that it can muster. Using the trickle of anaerobic metabolism, Loriciferans power their tiny bodies into places no other animal can stay for long.

A relative newcomer to zoological classification, *S. cinziae* has the potential to completely change our understanding of where animal life can be found on Earth. As one study puts it, 'a paradigm shift would be necessary, where textbooks would have to be rewritten'. But as the astrobiologist and science communicator Carl Sagan wrote in 1979, 'Extraordinary claims require extraordinary evidence.' And in 2014, one study struggled to find any evidence of living Loriciferans in L'Atalante and two other DHABs in the Mediterranean. Led by Joan Bernhard at Woods Hole Oceanographic Institution (WHOI), Massachusetts, just 16 specimens were discovered among 300 millilitres of sediment samples, a huge volume for animals that are often a tenth of a millimetre wide. Using a stain that glows when absorbed by living tissue, Bernhard and her colleagues didn't see the characteristic flash of life. Under a high-powered microscope, there were 'no identifiable internal organs in any of the loriciferans examined at high magnification', she wrote. Instead, they found scavenging bacteria and other microbes inside their hollowed-out husks, suggesting that the living Loriciferans that Danovaro and colleagues found were actually carcasses impersonating life. In a companion article to her study, Bernhard called them 'bodysnatchers'. They also found a Loriciferan in a sample of sediment taken from outside the DHAB basin, suggesting that they can live in places where there is oxygen available. To live in both places, Bernhard adds, is a level of resilience and ingenuity that seems very unlikely.

At least for an animal. For over two decades, Bernhard has studied a different type of anoxic basin, the Santa Barbara basin off the coast of California, a place where there are other life forms adapted to a life without oxygen. While there is no layer of salt, this bowl in the seafloor just happens to sit in a place with very poor circulation. The water in this basin becomes stagnant, the decomposition of matter that falls from above soaks up all the oxygen. Hydrogen sulphide leaches from the bacterial activity in the sediment. 'I work in stinky mud,' Bernhard tells me as we sit in her office at WHOI, her mouth covered by a white face mask. Then she tells me about her study subject: foraminifera, or forams for short. A single cell wrapped in a cloak of calcium carbonate, they are like amoeba with an ornate layer of armour. But it is their chemical ingenuity that makes them tough. When there is no oxygen they can metabolize nitrate, exhaling nitrogen gas. If there's no nitrate, there's a list of other compounds – manganese, iron, methane – that they might sustain their cell with (the studies are still under way). In a study published in 2021, Bernhard, her colleague Fatma Gomaa and their colleagues even found that they can digest hydrogen peroxide and liberate the oxygen within its chemical structure (H_2O_2). Even in anoxic waters, therefore, forams might be able to breathe oxygen.

These tricks have allowed forams to flourish. 'They live in the water column and in the mud, from the poles to the tropics, shallow water to the deepest abyss,' Bernhard says.

'So they're unbelievably successful. Plus,' because of their tough calcium carbonate shell, 'they have a fossil record. So that's how I got hooked on them in the first place. And now we know they can do anaerobic stuff, in addition to aerobic stuff.' In the anoxic Santa Barbara basin, Bernhard has found, there can be 300 individuals in a blob of sediment the size of a sugar cube, a density 20 times as high as well-aerated places. Bernhard's favourite foram, *Nonionella stella*, has even stolen the photosynthetic structures, known as chloroplasts, from the diatoms that live in the surrounding waters. What they are doing with light-sensitive organs[5] in the pitch black is just starting to be revealed. As Gomaa, Bernhard and their colleagues wrote in 2021, 'the potential metabolic capacity of foraminifera living in anoxic and hypoxic sediments remains largely uncharted territory'.

Anoxic waters have long been seen as uninteresting places where only stagnation and death thrive. The few scientists who studied these places were interested in arcane microbes such as bacteria or archaea. But this has changed over the last decade or so. Anoxic oceans, and the organisms that live there, are a window into the future. In 2017, a landmark study published in *Nature* found that since 1960, our oceans have lost 2 per cent of their oxygen. While this might not sound like much, the regions most affected by deoxygenation are often the most biodiverse and productive for fisheries. As one study puts it, 'no other environmental variable of such ecological importance… has changed so drastically and in such a short space of time'. Because warmer water can't hold as much oxygen (a result of the faster-moving water molecules bumping the gas back into the atmosphere), climate change has played a significant role in this widespread 'deoxygenation', particularly in places like the Arctic that are warming faster than anywhere else on Earth. But it isn't the only factor. Nutrient-rich waters from agricultural runoff or sewage outflows have increased the amount of growth in coastal regions, essentially fertilizing a dense mat of algae and other planktonic life forms with momentary excess. After they've consumed the nutrients and die, they are in turn consumed by oxygen-hungry decomposers, a legion of microbes that suck the breath out of their surroundings. Eutrophication, a process that essentially translates as 'well-nourished', is the precursor to a 'dead zone'.[6]

Perhaps the most famous 'dead zone' floats ominously off the coast of Louisiana and Tennessee in the southern United States, a strip of stagnant water that ebbs and flows, both in depth and length, with the seasonal outflow of the Mississippi River. With its tributaries draining 31 of the 48 contiguous, or 'lower', States, this enormous watershed is carrying three times as much nitrogen and twice as much phosphorus as prior to the 1950s, a direct result of the fertilizers sprayed onto vast monocultures of soy and corn that are grown to feed cattle and chickens or turned into biofuels for vehicle engines. This nutrient-rich runoff from farmland, combined with untreated manure, is flushed into the Gulf of Mexico where eutrophication takes over. In scientific papers, it is referred to as '*the* dead zone', an example to which all others are compared. At its peak, it is a blob of suffocating water the size of Wales in the UK (or New Jersey in the US), a deadly and invisible cloud that forces fish to either flee or asphyxiate, and kills worms and crustaceans that aren't fast enough to escape its descent.

A dead zone can come back to life, however. Even the one that sits in the Gulf of Mexico isn't static but shrinks in winter when nutrient outflows are at their lowest. And knowing that the primary cause of this nitrogen runoff comes from cattle and chicken farms provides a path towards recovery. As consumers, eating less meat not only reduces our emissions but also our impact on the water we depend on. In a distant and unexpected connection between our major sources of protein, reducing how much beef and chicken we eat can increase the populations of shrimp and fish we can catch along our shorelines.

In a more radical departure from our usual appreciation for nature, we might come to appreciate dead zones for what they are: ecosystems for life. Sure, larger, oxygen-dependent animals might suffocate and disappear, but these anoxic waters are fertile sites for single-celled protozoans such as forams. Speaking to Joan Bernhard in person and over video call, her enthusiasm for these creatures is infectious. Phrases like 'mind-blowing', 'very cute' and 'they're just awesome' come up several times in our conversations. Spending over 20 years studying these micro-organisms has only deepened her fascination with everything foraminifera. Their ingenuity and adaptability is all the more amazing given that they don't have a brain, she adds. What they do have is a cell full of chemical tricks and a means of sensing the world that seems almost human. As with an amoeba, forams can extend part of their cell into their surroundings, both as a way of pulling themselves from one place to another and also as a sensory exploration – a taste – of their local environment. Known as a pseudopod, this arm-like projection can reach far away from the shell, even rising from a place of anoxia to absorb the oxygen-enriched water above. Packed together in their breathless basin, I imagine *N. stella* reaching upwards in unison, a crowd of raised arms to say 'We are here, we always will be.'

A turtle's shell is a preadaptation to living without oxygen, a trait used for one thing (in this case, protection from predators) that was co-opted for another (anoxia). But it is not a prerequisite. When it comes to life without oxygen, there is one other animal that can compete with the painted turtle. It lives in freshwater lakes and ponds in Northern Europe, also battling ice and snow every winter. It might even have a representative in your house. Crucian carp, the drab, olive-green ancestors of goldfish, can survive for months without oxygen, a similar magic trick to the turtles but with a very different method. For one, crucian carp aren't air breathers. They pass water over their gills and extract the oxygen contained within. Compared to the anal sacs and skin of a turtle, it is a very efficient mechanism that can power an active life when oxygen is available. But when the snow falls and the sun disappears, they have no bony shell to protect themselves against the acidity of anaerobic metabolism. Instead, they have evolved the unique ability to turn lactic acid into ethanol, a highly diffusible molecule that they can excrete over the gills. This might not sound as wondrous as the secrets of a turtle's shell, but to a biochemist this is a marvel of evolution. As Sjannie Lefevre, a researcher at the University of Oslo, tells me, 'It's just astonishing that we have a fish that can actually produce ethanol.'

Carl Sayer, a pond restoration scientist based in southern England, is similarly astonished by these fish. Over his years breaking through bramble and willow to reach a bacterial sludge that was once a marl-pit pond used to irrigate farmland, he has learned to always cast his funnel-shaped fyke nets into the water, despite the seemingly inhospitable conditions. Pulling in these nets in the morning, he tells me, 'Everyone looks at each other and thinks, "No, there's no way." We think that it's almost impossible.' But, lo and behold, there are a few crucian carp flapping around in his nets. I ask if there are any memorable times when this has happened. 'A lot,' he says without the need to elaborate.

The few fish that remain in these forgotten ponds he calls 'survivors'. They are much larger than those that thrive in the productive ponds that are full of light, oxygen and life. While the latter may be the size of your finger, Sayer says, survivors may be as big as your hand. In some of the more isolated ponds that aren't connected by floods of high water, have no input or output of stream, he thinks that the fish might be the last remaining descendants of carp that were placed in this pond many decades ago for fishing or food.

While crucian carp can survive in anoxic ponds it isn't their preferred habitat. The large survivors don't have the resources to breed. They are holding out for better times. Pond restoration can bring back these productive ponds. Where there is light, plants can photosynthesize, releasing oxygen into the surrounding water. Where a lot of the sediment has been removed with a digger, there are fewer decomposing bacteria to suck up that oxygen and bury it. In such habitats, there can be hundreds of the smaller forms of crucian carp. The only limit to their survival, in these cases, is who their neighbours are. 'They're rubbish with predators,' Sayer says, 'like they can't get out of the way of them.' That's unusual, he adds. Coexistence is part and parcel of a functioning ecosystem. But crucian carp don't do well even when their fishy neighbours are just eating the same things as they do. 'It doesn't compete well with other species.'

When it comes to surviving abiotic extremes, crucian carp are the hardest of the hard. But when it comes to biotic influences – competition, predation – these fish are incredibly sensitive. This is a theme that comes up time and time again in those animals often labelled 'extremophiles'. In Chapter 6, we will meet heat-tolerant ants that are similarly sensitive to other species, whether predator or competitor. Entomologists, people who study insects, have a word for these species: subordinate. (Crucian carp researchers call them 'losers'.) If you put them in temperatures that are lethal to every other insect, they will thrive. Place them in a Petri dish with a member of another ant species, however, and they will be slaughtered. To survive the impossible is to leave yourself vulnerable to the mundane.

For a crucian carp, there are more immediate costs to life without oxygen. Ethanol, or C_2H_5OH, contains two carbon atoms, whereas the carbon dioxide, CO_2, that is normally exhaled when oxygen is available contains one. As carbon comes from our diet, in the form of sugars, this is like throwing away half of your food before you sit down to eat. As one paper noted, this way of dealing with lactic acid 'is wasteful of carbon'. Even so, the authors added sarcastically, 'this may be a relatively minor consideration for a vertebrate without oxygen'.

Another downside of living without oxygen, at least in crucian carp, is brain damage. Using a type of stain that highlights dead cells, Lefevre found that fish that had gone through anoxia showed much higher levels of apoptosis – programmed cell death. They also forgot how to navigate through a simple maze to reach a food item. To lose memories can be painful for us, as humans, to even comprehend. It is at the root of some of the most devastating diseases such as Alzheimer's or the side-effects of some of the most aggressive treatments for severe mental illness. For a crucian carp living in a cold lake, however, it's probably not that big a deal. They feel pain and deserve our respect as conscious beings but they don't have the same cognitive powers as we do. They certainly don't cherish the memories of their fry-hood, the birth of their thousands of offspring. For half of the year, their main concern is to keep their dorsal fin pointing upwards as the world above freezes. Plus, they have the ability to regrow the neurons that were lost. While it is a controversial topic in neuroscience with equally expert opinions saying that we can and can't regrow brain cells into adulthood, fish and other cold-blooded animals can. They may lose memories to the oxygenless winter, but they can recover and learn again.

Spring is a time for regrowth. Daffodils, apple blossom, plump green leaves full of chlorophyll, and, it turns out, the brain cells of crucian carp.

Before I left his office – and his turtles – in Toronto, Buck told me to set an early alarm for the next day. I was driving to meet Matt Pamenter, one of his former students who had set up his own lab at the University of Ottawa. A four-hour journey could easily turn into six or seven hours if I hit peak morning traffic, Buck warned. After grabbing a bite to eat, I settled into the room I had rented for the night, wrote up my notes from the day and set my alarm for 05:30.

As I followed the Ontario 407, heading northeast towards Quebec, the streams of red and white lights were already thick and threatened to jam. The skyline of the city, and its famous CN Tower, was still shrouded in darkness. Rain speckled the windshield. While the sun started to peek over the horizon, I began to wonder about all the lakes and ponds I was passing, many hidden among the dry stubble of maize stalks. Lake Scugog, Crowe Lake, Moira Lake and thousands of unnamed blue splotches on my map. Pumpkins lined the road outside farmhouses, preparations for Halloween in a couple of days. I imagined the painted turtles sensing the cool weather, the shorter days, their livers already bloated with glucose, and their sluggish attempts to build up their last reserves for the coming winter. Driving past the northern shores of Lake Ontario, I was just a few hours from the northern limit of this species, an invisible line that spans North America from Nova Scotia in the east to British Columbia in the west. Above snapping turtles, above red-eared sliders, it was a thick and dotted line that snaked above the Canadian border: *Chrysemys picta*, the painted, and most northerly, turtle.[7]

Avoiding the urban traffic, I made it to Pamenter's office in good time. Grey haired with a youthful face, Pamenter was dressed in a maroon hoodie, shorts and slip-on trainers, the casual sport-chic of the modern academic, the kind of outfit that's ideal for an informal Zoom call and an impromptu park run. Since starting his PhD in 2003, Pamenter has studied how the brains of turtles cope with anoxia, collaborating with Buck for many of his projects. For the last decade or so, he has drifted away from overwintering reptiles of North America and focused his attention on a bizarre mammal that lives in the arid Horn of Africa. After a brief chat in his office, he leads me to the room where they are kept. Compared to a terrarium of turtles, Pamenter's room is much warmer and a heck of a lot smellier. Instead of trickling water and the hum of filtration systems, there is a constant soundtrack of scratching and squeaking. After a few studies found naked mole-rats to be hypoxia-tolerant, Pamenter joined a growing cadre of scientists who study these wrinkly, Mars bar-sized rodents with huge, goofy incisors, a famously ugly mammal that has been likened to a penis with teeth or a sabre-toothed sausage.

From pictures, they certainly fit their description. Meeting them in person, however, I thought they were adorable, like day-old puppies with oversized gnashers. Removing his hoodie and donning light blue surgical gloves, Pamenter opens one of the plastic cages that were made for laboratory rats but had been adapted for their much smaller and far more social relatives. Each chamber is connected to another by a short piece of plastic piping, a simulacrum of the underground tunnels these rodents excavate in the dry soils of Somalia, Ethiopia and Kenya. Forming colonies of up to 300 individuals, naked mole-rats are eusocial: like termites, bees, wasps and ants, they are ruled by a dominant female – the queen – and allocate breeding to just a few chosen males, while the rest of the colony maintain their nest or care for the young.[8]

As we shall see in Chapter 6 on heat-tolerant ants, eusocial species are often able to conquer the harshest and most unpredictable environments, their large number of foragers able to locate sparse and infrequent food sources. A sterile forager is more expendable – evolutionarily speaking – than a breeding female. If an ant is taken by a lizard, or a naked mole-rat by a snake or bird of prey, the future of reproduction is little affected. In the arid soils of East Africa, digging is impossible for much of the year, the parched soil turned impenetrable to even the sharpest incisors. Whenever rain falls, there is a brief window of opportunity to extend the maze of tunnels in the search for the enormous, cassava-like tuber of *Pyrenacantha* or the smaller roots of the *Macrotyloma* bean. Having dozens of workers all digging at once can mean the difference between finding food and not, between an animal living and dying. As one study puts it, 'the animals must cooperate and dig furiously to locate food patches'. Once located, a naked mole-rat informs her nestmates with repetitive chirps that have been called 'recruitment calls', just one type of vocalization this species uses among a repertoire of 'high-pitched contact and aggressive chirps, a mating call, toilet-assembly call, and vocalizations specific to pups, such as squawks when pups are stepped on and caecotroph-solicitation chirps' – that is, 'can I eat your poo?' requests. All together, another paper boldly claims that 'the current data suggest that the naked mole-rat is one of the most vocally complex rodent species, with a repertoire that even rivals that of primates.'

To take advantage of a particularly bountiful food source, ants follow the thickest trail of pheromones left by their colony, honeybees translate the 'waggle dance' of a returning forager into distance and direction from the nest, and naked mole-rats excitedly sing to each other in complete darkness.

Eusociality, it turns out, might be a beneficial strategy across much of Africa in the not-too-distant future. 'Everything's going to become hotter and drier,' says Hana Merchant, a researcher who studied the common (non-social) mole-rat for her PhD in the greener, more Mediterranean-style habitats in Central Africa. 'Potentially, we'll get even more hyper-arid conditions that don't exist at the moment.' With climate change making soils hotter and drier, reducing the amount of food available, the naked mole-rat might be a window into the future – for mole-rats, at least. Since eusociality is also found in the Damaraland mole-rat, a species whose colonies are smaller (around 40 individuals) and ruled by a king (not a queen), this group of rodents has an innate desire to cooperate when times – and soils – get tough.

Harsh environments often mould hardy inhabitants. And naked mole-rats are certainly an example of this.

In Pamenter's room, there are 300 naked mole-rats separated into six colonies, each with their own shelf on an industrial-looking shelving unit. Each chamber is a riot of activity. In one chamber, among a mass of writhing bodies, Pamenter points out the queen. Although noticeably larger than the others, it is her shape that is most distinctive. Able to birth 20 pups at a time,[9] her belly bulges to either side of her like a pair of overstuffed panniers on a road bike. To accommodate such a load, a queen lengthens and fuses her vertebrae. This happens for every female that rises to monarch. In this mammalian society, being queen isn't a case of wearing a crown and living in a palace. Royalty is in your bones.

If there is a comparison to human monarchs, the naked mole-rat queen would be a mediaeval autocrat, one who doesn't wave at her followers but rules with an iron fist. For decades, it was thought that some unknown pheromone must keep her nestmates in line, a waft of chemically induced subordination. In reality, she bites, nudges and squishes to assert her dominance. If there's a particularly busy tunnel in her colony, the queen will always walk on top of the pile.

Pamenter leaves the queen where she is. Picking her up might disturb the colony and a number of the queens in this room are pregnant. It also seems hugely disrespectful to suddenly hijack the most powerful animal in their world. Instead, he picks up the second largest individual, one of the breeding males. Although they are rodents, they seem more docile than a rat, perhaps even more than a pet hamster. In ten years of working with these animals, Pamenter tells me, he has been bitten only once. I marvel at how delicate this animal looks, pale and wrinkly, its eyes just poppy seeds without expression. But this puppyish appearance is deceptive. Naked mole-rats are undoubtedly some of the hardiest animals on Earth. They are resistant to cancer. They've lost many of the receptors that signal pain. And while a similarly sized mouse will live two or three years, a naked mole-rat can live to well into its thirties.

This resilience and extraordinary longevity have made these animals into the latest biomedical models of ageing and cancer.[10] It wasn't until 2009, a time when Pamenter was working in California, when his interest was piqued, shifting away from his research into painted turtles. That year, two studies were published that tested naked mole-rats' response to hypoxia. Brain slices taken from the rodents seemed unperturbed by oxygen levels reduced to 10 per cent. Their neurons still fired with action potentials as if they were in the atmospheric 21 per cent. The activity of brain tissue taken from a mouse, however, declined by half. They were struggling to function. Whether oxygen was reduced to 10 per cent or all the way down to zero, the naked mole-rat tissue maintained its activity for longer and was able to recover its function when oxygen was reintroduced. In comparison, the authors wrote, 'the mouse [tissue] did not recover at all'. Then, in 2017, another study found that naked mole-rats could survive complete anoxia for 18 to 20 minutes, five times longer than the similarly sized mouse. After this period, they returned to their nestmates, perhaps a little disgruntled but nonetheless untroubled.

Why would a mole-rat have such an indifference to anoxia? The most logical – and leading – theory was rooted in their subterranean lifestyle. Spending all their time in underground burrows has long been thought of as an oxygen-poor existence, especially during the wet season when moist soil acts as a barrier to ventilation. And naked mole-rats have an additional contributor to asphyxiation: they form large social groups. With some colonies around 300 strong, each mole-rat absorbing oxygen and exhaling CO_2, it was a solid assumption that their nests would be very low in oxygen. The only problem is that there's no evidence that this is actually the case. No one had been able to provide a reliable measure of oxygen levels from their natural burrows in East Africa. I found this extraordinary. For all the modern experiments that can measure activity of a single receptor within a neuron, the release of certain ions of sodium and potassium, how hard could this be? As we sat in his office, his desk facing a wall of family photos and a clock with a naked mole-rat printed onto its face, Pamenter tells me, very hard. Naked mole-rats are incredibly sensitive to sound, he says. The footfall of a harmless scientist is a cause for alarm. Such a scientist could drop an oxygen probe into a naked mole-rat burrow no problem. There just won't be any naked mole-rats near it. Some nests have been found to extend for four kilometres and they will move around this labyrinth and sometimes never return. Jane Reznick, a naked mole-rat researcher working in Germany, tells me about one scientist in the 1960s or 1970s who left his probes in the ground, hoping that they would return once he had left. And they did. But as they came across the probes, they chewed them into disrepair.

Even without an accurate measurement from the wild, Reznick is pretty sure naked mole-rats are, at least at times, experiencing very low levels of oxygen. In her lab in Cologne, she has a range of nest chambers and pipes of different diameters. Without exception, her colony will sleep in one huge pile-on, a wrinkly mass of bodies, and choose the smallest chamber. 'It's very cute,' she says. 'But the animals on the bottom must be getting very little oxygen.'

To further investigate the reason for the naked mole-rat's extraordinary tolerance to low oxygen, Pamenter travelled to South Africa in 2019. While naked mole-rats live in the arid horn of the continent, they have relatives that live in more temperate regions. They are larger and hairier but not as social as their naked cousin. Most mole-rat species are solitary. It was this natural diversity – animals living alone or in hundreds – that Pamenter wanted to make use of. By testing their ability to withstand hypoxia, he could then see whether it was the solitary or eusocial animals that differed most when it came to surviving in low oxygen environments. In a study published in 2021, he found that this wasn't the case. While the naked mole-rat was undoubtedly the champion of hypoxia, all species tested were still far more resilient than a laboratory mouse. Whether they were solitary or eusocial, living in low oxygen didn't seem to be a major problem for a mole-rat.

Bedtime pile-ons might be adding to a naked mole-rat's exceptional hypoxia-tolerance, but it's not the main reason for it to evolve. Plus, subterranean lifestyles are found across the mammalian tree: moles, golden moles (actually related to elephants), rabbits, tenrecs. Similar in lifestyle, none of these animals tolerate hypoxia like mole-rats do. But, on closer inspection, none of them dig like mole-rats either. They all dig with powerful forefeet. Mole-rats, meanwhile, dig with their mouths. (The phrase scientists use is 'teeth and head-lift diggers'.) A quarter of their entire body's muscle mass is used to open and close their jaws. Using their huge, self-sharpening incisors, they munch through compacted soil like a sewer rat might chew through a drainpipe. When it comes to their name, mole-rat, the emphasis is on the rat. The mole is their way of life – subterranean. The rat is their physiology. When they dig, they get a face-full of dirt. In captivity, mole-rats hold a piece of wood chip behind their incisors to prevent soil from going down their throat and choking them. Perhaps this is the reason all mole-rats seem to tolerate low oxygen environments. You can't breathe when you are digging with your face.

Although naked mole-rats can survive anoxia for a time, the more ecologically relevant research asked the question of how they live happily in hypoxia, levels of oxygen around 7 per cent. This chronic level of oxygen deprivation might reveal how certain disorders such as chronic obstructive pulmonary disease may be eased over many years of living with the condition, Pamenter hopes. He admits that he doesn't know how their abilities could be translated into a medical treatment. What gives him hope is that discoveries come from the most unlikely places. The naked mole-rats' resistance to cancer, for example, comes from a protein that prevents cells from sticking together. Produced in abundance in the skin of these wrinkly mammals, it prevents them from getting snagged or stuck in their labyrinth of tunnels. 'So, stretchy skin for life underground leads to cancer resistance, contributes to longevity,' Pamenter summarizes as he leans back in his desk chair. 'You'd never see that coming.'

The snow melts into mush before the ice flows back into water. The turtle at the bottom of the pond, positioned in the shallows, can sense the change. Light. A slight tingle of warmth on her thick skin. Her heart starts to beat faster, a beat every minute instead of every ten. The duckweed and algae start to grow once more, releasing oxygen into the water; the first fresh bubbles since the snow fell in November. In a few days, the sun has reduced the ice to a thin glaze and holes appear around the stalks of vegetation from the previous year. The turtle's legs are stiff from inaction, heavy from the cold. Her lungs have been collapsed for the entire period of dormancy and they provide no buoyancy. She has to force herself to the surface, the last dregs of glucose in her liver being released for the effort. One kick with her left claw, then her right, she rocks upward, a puppet on a string. Exhausted, her tubular nostrils poke through the crumpled ceiling of her former confines and she inhales for the first time in months.

This first breath of oxygen is like a taste of life itself, a sensory throwback to when anaerobic microbes evolved into complex, multicellular animals. Her muscles, neurons and every fibre of her being are supercharged by the return of this combustible gas, a suite of antioxidants mopping up the damage and preventing serious reperfusion injury. The lactic acid that has built up in her blood, the excesses that weren't buffered by the shell, is converted back to glucose and stored in the liver. And every part of this process is a tiny chapter in a much larger story: one of survival in places where other animals would die. This particular painted turtle, able to endure months with oxygen, has first dibs on the coming spring. As the world around her warms, just a little, she can forage for plant matter without competition, gorge with little fear of predation. 'If you can survive that seasonal ice cover,' Les Buck tells me in Toronto, 'tough it out for four months, then you have a niche all to yourself.'

We generally think of nature being 'red in tooth and claw', a constant whir of predation and competition. Sometimes it is simply finding a place in space and time where you can be left alone. A painted turtle in a freshly thawed pond, a crucian carp in a stagnant hollow in the ground, a Loriciferan in a brine-filled basin: animals can be extroverted in their survival, happiest in those places that exclude others.

CHAPTER THREE

FASTING & FURIOUS

Food

On 29 December 1946, while hiking through a narrow canyon in the Chuckwalla Mountains of California, Edmund Jaeger spotted a bird. At least, he thought that's what it was. A common poorwill, a nocturnal bird with a mottled grey plumage, no bigger than a starling.[1] But it wasn't flying through the air above, wasn't perched on one of the shrubs

that took root in the steep rock, and it certainly wasn't singing. It was nestled into one of the crevices of cold hard granite, as if the mountain had accepted this creature as part of its own topology. Its plumage was almost indistinguishable from the rock it clung to. Like a limpet clinging to a boulder at the beach, it was fixed and unmoving. Accompanied by two of his students from Riverside College in California, Jaeger took a closer look and decided to give it a poke. 'I even stroked the back feathers without noticing the slightest movement,' he later wrote. As questions raced through his mind about what it was doing there – was it dead? Frozen? Sick? Lost? – he carried on walking up the path to the top of the mountain.

He and his students returned a couple of hours later. The bird was still there – same position, same indifference. Jaeger decided to pick it up, gently caressing it from its crevice like picking a pistachio from its shell. He turned the bird about in his hands and saw it was a common poorwill, but it felt lighter than the other birds he had held in the past. Its feet and eyelids were cold to the touch. Without any sign of life, not even a sigh of sleep, he decided that care and quietude were no longer needed. 'We made no further attempt to be quiet,' he wrote, 'we even shouted to see if we could arouse our avian "sleeper."' After he returned the bird to its crevice, it lazily opened an eye – 'the only sign I had that it was a living bird,' Jaeger wrote.

The common poorwill (*Phalaenoptilus nuttallii*) is a species common to arid and semi-arid environments of North America. A nocturnal insectivore that catches its prey – beetles, flies, moths – on the wing (and in the dark), its call was well known to anyone who lived in this region – *Poor-will-ip*.

Poor-will-ip. Poor-will. 'Their voices… lend much to the fascination, for the notes seem to be part of the night itself,' Arthur Cleveland Bent wrote in his surprisingly readable 1940 book, *Life Histories of North American Cuckoos, Goatsuckers, Hummingbirds and Their Allies.* '[T]he bodies of the birds themselves seem more like detached and living particles of darkness than of flesh and feathers… it does not seem a bird at all, but simply "a wandering voice".' Also known as goatsuckers for the mistaken belief that they would sneak into a shepherd's hut to raid the udders of their flocks,[2] common poorwills were thought to migrate south, Jaeger knew, following the availability of food into southern Texas and Mexico. When insects disappeared from the desert, so did the night's wandering voice.

What was this bird doing here? Did it hold onto the memories of summer a little too long, anticipate that one insect too many?

Jaeger returned ten days later, walking up the mountain under a grey sky. Not only was the bird still there, it hadn't moved 'even so much as a feather'. This time, however, its breath was perceptible and it produced some squeaking noises like a mouse. After what looked like a yawn, it then 'resumed its quiet'. In his writings, Jaeger called it a 'comatose condition' and suggested that the common poorwill didn't migrate to warmer climes in winter. It hibernated. While this phenomenon was very familiar to scientists, this was a huge claim. Brown bears, ground squirrels, marmots: they all hibernate through the hardships of winter. But they are all mammals. There were no known hibernating birds. Out of over 6,000 species belonging to Class Aves, the common poorwill would be the first.

Jaeger returned to the Chuckwalla Mountains a year later, in November 1947. He found the same crevice filled with, presumably, the same bird. As hibernation is a state of torpor that lasts for more than one day, he then monitored the bird's temperature and movements for the next four months, returning most weekends to continue his research. Its body temperature varied little, hovering between 18 °C and 20 °C, even when the air warmed up to 24 °C during the day. A medical stethoscope couldn't detect any heartbeat. A mirror held in front of the nostrils collected no moisture, no breath. 'No movement of the chest could be detected,' Jaeger wrote in the journal *Condor*. 'All this together with the low internal temperature I take as evidence that the bird was in an exceedingly low state of metabolism, akin, if not actually identical with hibernation as seen in mammals.'

He then lifted up one of the bird's eyelids and pointed a small torch directly into the pupil. 'To this strong stimulation there was no response whatever, not even an attempt to close the eyelid,' he wrote, 'a remarkable evidence of obliviousness on the part of the bird to what was going on in its environment.' It was similarly oblivious to a harsh storm that swept through the mountains on 7 December, tearing its tail feathers with an onslaught of hail and sleet. A layer of ice covered the ground after it had passed. The bird remained, battered but 'unaware of [the storm's] fury'.

Although new to science, the poorwill's ability to hibernate through winter had been known to others for centuries. The Hopi tribe of this area call the bird Hölchko, 'the sleeping one'. When Jaeger asked a student who was a member of the Navajo tribe where this bird went in the winter, he replied without hesitation: 'Up in the rocks.' Taking this as corroboration of his own observations, Jaeger wrote that 'Poor-wills are rock seeking, hibernating birds in winter… The utility of such hibernating birds is obvious. During the period when there is little food accessible, this bird, instead of migrating to areas where insect food is available at night, may go into a state of inactivity.' When he lit his campfires against the cold of the December night, Jaeger didn't see any moths or other insects fluttering through his camp. It was a place of cold famine.

As we shall see, other birds travel vast distances to feast on their preferred diets, a strategy that has allowed them to fly from one summer to another, in a matter of days. One such bird, the bar-tailed godwit, flies non-stop over the Pacific Ocean, from the Arctic to New Zealand in just over a week, a trans-oceanic (and trans-hemispheric) migration that's powered by worms and clams that breed in warm mudflats.

Long-distance migration avoids the inhospitable. Hibernation braves it.

It was a lack of food, not a drop in temperature, that was vital to the poorwill's hibernation. This was demonstrated most convincingly by Joe Marshall Jr, a zoologist, ornithologist, talented watercolourist and harpsichord enthusiast who had also served as a parasitologist on the island of Saipan (part of the Empire of Japan until July 1944). Marshall Jr's dedication to biology is clearly demonstrated during his time in a war zone. While a Marine sergeant was scouring the cliffs of Saipan for snipers, he stumbled into a comrade holding a very large net. As he was positioned here, Marshall Jr thought it was a good opportunity to sample some of the swiftlets that build 'edible nests' on these shores, hence the net. After returning to the US, he completed his PhD in California and published papers on local mammals and birds. And in the autumn of 1950 he collected a few common poorwills and kept them in his shed.

Daylight shone through a Plexiglas wall and illuminated a sand-covered floor, a few shelves, and a sack of chicken feed in the corner. In this humble setting, Marshall Jr set up his experiment, controlling what these birds ate, and when they ate it.

From October to December, he provided moths that he attracted with a torch and gave to the birds by hand. On 28 December, the first heavy frost of the year, he withdrew food entirely from one of the birds. While the unfed bird became torpid, tucking its beak into its chest and taking on the 'comatose condition' that Jaeger had seen in the wild, the two birds that were fed 'remained active no matter how cold the weather… Low temperature therefore appeared not to induce torpor directly, although it helped maintain that state once it began.' It was only until the bird had lost around 20 per cent of its body weight that it entered the state of hibernation. Emerging from their torpor in March, the birds were immediately the antithesis of dormant. 'They awoke in a snarling and voracious condition, as if forced to heightened activity to prevent starvation.'

Hibernation is rare in birds but common in mammals. It has allowed the boreoeutherians ('northern true beasts')[3] to reach into seasonally cold and warm places, the northern latitudes. When times are cold and food runs low, they can depress their metabolism to conserve their energy for better times. As we shall see in the chapter on freezing, ground squirrels that live in the Arctic have become a model system for how mammals hibernate. But this choice has also blinded us to the true nature of hibernation. Even tropical mammals can enter this coordinated state of low metabolism, without ever experiencing the cold. In 2004, a lemur was discovered to spend several months in a nest hole, its body temperature shifting between 9°C and 35°C. Below these arboreal lemurs, tenrecs spend nine months of the year hibernating in burrows, a response that Frank van Breukelen, a researcher at the University of Nevada, Las Vegas, who studies mammalian hibernation, thinks is actually a response to high predation risk at certain times of year. 'Before [studying] the tenrecs, I would have told you it was all about temperature,' van Breukelen tells me. 'I really believed that animals needed a low temperature in order to depress all these metabolic processes and save themselves all the energy.' Now, he adds, 'they've confused the living crap out of me and I have no idea anymore what they're doing.'

Even the word hibernation comes from the Latin *hibernātus*, meaning 'winter'. This presumption is forgivable. The boreoeutherians are by far the most common group of mammals – comprising 6,000 of the total 6,400 species – alive today. Remove this group from the family tree and you'd be left with a few armadillos, elephants and manatees, marsupials and monotremes (platypuses and echidnas). And then there's the bias that comes from where science originated, largely in northern, mid-latitude regions of Europe and North America. To take a hibernating ground squirrel as the norm for hibernation is as understandable as it was wrong. Since mammals evolved in a tropical world, one without glaciers and ice poles, the ancestral hibernator would have been more like a tenrec than a ground squirrel.

Hibernation in a warm climate may even have allowed mammals to survive under the feet of dinosaurs, a time when temperatures were rarely below freezing but the predator risk was – to put it mildly – very intense. Then, when a big asteroid crashed into modern-day Mexico, their ability to ramp down their metabolism, despite the fires raging above ground, may have been the key to their success. The end of the dinosaurs was the beginning of the age of mammals, a period in Earth's history that has given rise to a super-intelligent primate that walks on two legs – us.

Mass extinction is, by definition, largely a phenomenon of dying. And it's not just the death of individuals but entire groups of animals. Shapes, sizes and modes of living, no matter how long they have been present on our planet, can blink out of existence. Adaptations that may have served an animal well for millions of years prove to be unfit for a dramatic shift in environment, whether it comes from an asteroid impact, widespread volcanic eruptions or climate change. The unpredictable nature of catastrophe can completely reshape Earth's inhabitants. Long before the Cretaceous extinction, for example, there was the Permian extinction, a time 252 million years ago where widespread volcanic activity in what is now Siberia caused extensive anoxia in our oceans, killing 96 per cent of life.[4] With 70 per cent of all life on land also falling into extinction (including a few groups of insects that rarely succumb en masse), the end-Permian extinction has been called the 'Great Dying'.

First noted from the fossil record in 1841, modern palaeontologists define this moment as 'a time of extraordinary renewal and novelty'. Waters dominated by filter-feeding animals attached to the seabed were replaced by a fauna that was more predatory, active, able to swim and sense their way through the world. By clearing the dominant rugose corals, brachiopods, trilobites and crinoids, a new world of predatory cephalopods, crabs, snails, sharks, bony fish and marine reptiles could emerge. As one paper put it in 2022, these animals 'were all faster and nastier' than their Permian precursors. This is the world we still live in today. The extraordinary diversity around us – shell-piercing cone snails, sentient octopuses, eight-metre-long great white sharks, toothed whales hunting giant squid in darkness – has its roots in the greatest catastrophe life on Earth has ever experienced.

Whether it's due to food shortages or simply clearing an ecosystem of predators, mass extinctions are periods of immense opportunity; a slate that hasn't been wiped clean but has significant space for new sketches. At the end of the Cretaceous, some 66 million years ago, it was a series of inhospitable conditions that put an end to the dinosaurs' reign. The asteroid led to burning hot showers of glass and rock, increasing the air temperature to such a degree that fires spontaneously ignited among the conifers and ginkgo forests. The cloud of debris from the impact crater blocked out the sun and prevented plants from photosynthesis, crippling ecosystems at their foundation. Without photosynthesis, the oceans and lakes lost their oxygen. Without sun, burning heat rocked quickly into freezing cold. Dinosaurs, pterosaurs, marine reptiles such as pliosaurs and ichthyosaurs, and shelled ammonites all succumbed. In their wake, the subordinate became the dominant force, the subterranean furballs evolved into fast-running predators on four legs, giant whales grew larger than any dinosaur ever could, and skin-winged bats flew where pterosaurs once soared.

To sterilize a living world is near impossible. When seen over geological time, the destruction of life is merely the creation of something new.

Every mammal that hibernates – whether under the feet of dinosaurs or in the warm soils of modern-day Madagascar – is, for a time, reverting to a more reptilian mode of life. Just as a turtle endures the anoxic waters of a snow-covered pond, metabolism is ramped down to conserve the resources built up over a previous season of plenty. A formerly warm-blooded animal becomes cool. This trick has allowed ground squirrels to endure the harshest Alaskan winter, cooling their bodies to −3°C for several months every year. To survive the cold, they become cold-blooded. It is a tried and tested solution to surviving in places with unpredictable or infrequent food. Long before mammals conquered the northerly climes, however, snakes, crocodiles and lizards were performing such metabolic depression to live in places that few larger animals could endure.

Less than a square kilometre in area,[5] Shedao Island is a tiny chunk of sandstone that emerges from the Bohai Sea off the coast of Liaodong Peninsula, northeastern China. Shedao is too small and too meagre for most animals to eke out a living. Its name translates as Snake Island for a very obvious reason: there are 20,000 snakes here, all *Gloydius shendaoensis* – the Shedao pitviper. The only other vertebrates that inhabit the island are a species of bat, two seabirds and northern white-rumped swifts, all of which can fly the 11 kilometres to the mainland if food runs low.[6] The metre-long snakes that curl along branches, their patchwork of light and dark grey scales merging into the canopy's shadow, cannot. They are residents here. But they wouldn't leave even if they could; for six weeks every spring and autumn, flocks of migrating songbirds settle on the island, resting and feeding among the shrubs and trees. Emerging from their burrows and crevices in the sandstone canyons, *G. shendaoensis* slither onto their favourite branches, retract their heads to form a spring from their forebody, and then wait for a bird to alight on its perch. Preferring the snack-sized warblers and buntings, these snakes can consume a quail if the opportunity arises.

After a few weeks of ambush feeding, the snakes return to their dens and remain inactive for half of the year. It's an impressive period of fasting but not extraordinary for a snake. A python, pregnant with its clutch of eggs, can fast for 18 months. In captivity, some snakes, which I can only imagine were incredibly hangry at their owners, didn't eat for over two years. While keeping completely still, a snake's physiology transforms.

The secrets of fasting snakes are not only a product of their low metabolic needs but also what they do with their digestive systems. Although intestines aren't oxygen-sinks like the brain or heart, they are still energetically expensive. The cells that absorb the nutrients from the food we eat are being replaced all the time. But while our digestive systems are in near-constant use – swallowing food at the mouth, breaking down food in the churning stomach full of acid and massaging it through the looping intestines – snakes shut their digestion down completely. For them, the phrase 'use it or lose it' is a gut reaction. The intestines shrink, the transport of nutrients ceases, and the snake diverts its blood flow to other parts of the body.

When a snake catches that next meal, the digestive system has to be rebuilt and switched back on. To power this transformation, the metabolic rate of a Burmese python can increase by as much as 40 times and remain at this hyperactive state for two weeks. (After a large meal, our metabolism will increase by half compared to before we started eating.) And this is before the python has absorbed any sugars or lipids from their prey. It was still in their throats. This burst of energy came from its stored reserves which rebuilt the digestive system in just 12 hours. (Snakes also rebuild their hearts after months of relative inaction.) To kickstart a whole organ system takes a lot of energy. Roughly a quarter of all the energy the snake would get from its prey would be used to replenish used energy stores of fat and protein. But, the researchers who studied this response to fasting write, this strategy is still cheaper than 'maintaining the intestine uselessly ready for months'.

Shedao viper or a captive python, to be a snake is to be starved of food. As a review paper from 2016 puts it, these legless reptiles 'exist predominantly in a fasted state'. Hunger is the norm, satiation the exception.

A polar bear is perhaps the least snake-like animal you can think of. Round, furry, big powerful limbs, dinner plate-sized paws. And yet they endure a similar, almost reptilian, period of fasting for four to eight months every year. While males maintain their activity levels throughout this period, wandering for thousands of kilometres every year, the females dig a den for their cubs, feeding them with milk that contains 30 per cent fat. Eating snow to maintain her fluid balance, a female polar bear can lose half her body weight every year; a 300-kilogram pregnant female becomes a 160-kilogram single mother of two.[7] As the long winter breaks into spring, a starving mother bear has to begin her feast on the fattiest of foods: seal blubber, a buffet that allows the largest land predator to survive in an otherwise meagre world of ice and rock.[8]

There are eight species of bear alive today,[9] and the polar bear might be the most atypical of them all. (The herbivorous, bamboo-eating panda is also a strong contender.) While the other members of the Ursidae clan are all land animals, *Ursus maritimus* is a marine mammal, living most of its life on ice floes and swimming great distances to hunt. The largest grizzly and the largest polar bear might look similar, a few kilograms between them and a white covering of fur instead of the ancestral brown. It is where they live and what they eat that make them incomparable. 'It is a novel niche,' says Andrew Derocher, a polar bear researcher who has studied these animals for over 40 years. 'Because the ringed seal, the main prey of polar bears, has been around for a lot longer than polar bears have, they must have had really happy days before the bears started to, you know, show up and start to eat them.' Seals would have been hunted by killer whales before this, but adding a polar bear on their ice floes just seems like really bad luck.

This is what evolution does to you when you envelop yourself in a five centimetre layer of fat. A harbour seal from colder climes might be seen as an appetizing sausage for the ocean's largest predators. A seal living in the Arctic, however, is life's version of a pig in a blanket. If there is a reason polar bears endure freezing cold winters, four months of constant darkness every year and long periods without eating, it's because they discovered a buffet of unimaginable calories. It is an obsession with blubber.

From an ancestor that ate caribou and foraged for berries, polar bears have specialized their being to eating seals. Their white fur camouflages their large bodies as they approach a seal's breathing hole. Their flattened skulls and upward-facing eyes allow them to swim and still breathe and observe. But it is inside their body where their seal-specialization is most pronounced. While snakes will convert 80 per cent of their prey into their own tissues, known as the assimilation efficiency, polar bears turn 90 per cent of seal blubber into bear fat. 'That's what makes it possible to be a polar bear,' Derocher says. 'They go around, they kill the seals, they peel them open, suck the fat, put it on their body, and then they just keep going. They're on this tier of energy intake that is just off the charts.' In lectures to the public, he calls polar bears 'fat vacuums'.

The ringed seal is what keeps these bears going. But they also tackle bigger prey if the opportunity rises. Bearded seals, beluga whales trapped between ice floes, walrus. Each time, the bear can eat around 20 per cent of its body weight in one sitting. A 500-kilogram polar bear, that is, could suck up 100 kilograms of fat from a bearded seal, 90 kilograms of which is going straight onto its body. 'We've had polar bears go from 98 kilos one fall to over 425 kilos the next year,' says Derocher. 'So these are animals that go through these massive, massive swings in body condition.'

Along with carbohydrates and proteins, fats are a main constituent of life on Earth. In our human culture, this family of molecules is often seen through a negative lens. To be flabby, overweight or obese is stigmatized while the lean and muscular are viewed as ideals. From a purely medical point of view, obesity and high fat diets are associated with poorer health outcomes such as heart disease, diabetes and fertility problems. But wild animals don't seem to struggle with such consequences, even when they are at their fattest. For marine mammals like seals or baleen whales, this is partly due to the position of the fat cells; by depositing their fat in a layer under the skin, the organs and muscles are free to flex and communicate. Fatty deposits aren't a problem when they are deposited safely. In obese humans, fat is stored throughout the body (known as visceral fat), surrounding the heart, lungs and blood vessels in a restrictive coating that makes each organ have to work harder. Locomotion, such as walking, is also impeded. But a bowhead whale that has a layer of blubber between 20 and 40 centimetres thick can migrate 1,500 kilometres every year, and lives for over 200 years.

In nature, obesity can be a vital means of surviving in places where food isn't always available. Even when resources are year-round, near the tropics for example, competition between animals might be so intense that consuming as much as you can eat when you can is a wholesome strategy. Overeating, or 'hyperphagia', provides an animal's body with enough reserves to get through inevitable lean times. For a polar bear, periods of extreme fasting are due to seasonal changes in the same place. Similarly, emperor penguins in Antarctica can fast for months, remaining largely inactive until they can return to the sea to feed. Whether mammal or bird, the process of fasting proceeds through three stages. In the first days without food, fat reserves begin to be utilized and the body mass of the animal experiences its first decline. The second phase can last for weeks and involves the extensive use of fat and protein stores, both being converted into glucose to power the body's cells. (A king penguin chick can remain in this stage of fasting for four months.) The third phase is defined by a near-total loss of fat from lipid deposits and the body starts to digest lean tissue such as muscle and organs. Body mass plummets in this final stage and nutrients are exhausted. Although not officially used in the scientific literature, the fourth phase is either refeeding or death.

While the body is consumed to power itself, it also pays to reduce energy expenditure. A king penguin that decides to waddle and hop over the ice floes during its seasonal fast would die much quicker than its neighbour who stood still. A polar bear that takes regular breaks to nap survives longer than one that walks endlessly for the four months of winter dark. But this doesn't mean that fasting is synonymous with inactivity. Elephant seals that haul themselves up onto the Californian coast don't eat for up to nine weeks while defending their harem and mating (males), or spend four weeks converting their fatty reserves into similarly fatty milk (females). With no access to food for nine weeks, blubber is the lifeline that keeps them going. 'Blubber has always had that assumption that it's just there really to keep the animal warm,' says Kimberley Bennett, a scientist from Abertay University in Scotland who studies the blubber of seals and its influence on the physiology of these animals. 'But most marine mammals have extended periods of time where they've moved away from their food supply because they're doing something else, like breeding. Or it'll be a year when there just isn't really a regular food supply. So they kind of need that backup.' Male elephant seals may lose over a third of their body weight as they actively defend their harem of females. (For an animal that can weigh three tonnes, this can be a literal tonne of weight loss.) Once the breeding season is over, the adults return to the sea to replenish their reserves. But the young pups are left without sustenance after they wean; they might not eat again for another four months. Elephant seals, perhaps more than any other animal, are brought into a world defined by food's absence.

Whether we travel back to our origins in the Rift Valley of East Africa or our early peregrinations around the planet,

Homo sapiens has long had to face periods of food shortage. Whether hunting antelope or foraging for berries, seasonal shifts in prey or flowering plants would have led to periods where fasting was inevitable. The further north our ancestors roamed, the more extreme the winters became. As a 2019 review in the *New England Journal of Medicine* states, 'our human ancestors did not consume three regularly spaced, large meals, plus snacks, every day, nor did they live a sedentary lifestyle'. To fast is to reconnect with our past, the authors argue.

But the connection is deeper than this. Fasting is a biological response to reality shared by the majority of animal life. The naked mole-rats we met in Chapter 2 have limited food supplies but live extraordinarily long lives free of cancer. A cave salamander that lives under the Dinaric Alps, Southeastern Europe, doesn't have to eat for a decade but can live for over 100 years. A 40-year-old rhesus monkey that has eaten 30 per cent fewer calories than his laboratory mates appears as young as monkeys half his age, his hair still thick and ginger, his skin not saggy, and showing no sign of cancer or other age-related disease.[10] Studies in humans and animal models such as mice and rhesus macaques have shown that reducing the number of calories eaten, or extending the number of hours between meals, can increase lifespan while reducing age-related diseases such as cancer, diabetes and dementia.

Whether fasting for a couple of days a week or reducing the number of calories consumed by 20 to 40 per cent, the same biological process unfolds inside the bodies of animals, including ourselves. A few hours after beginning to fast, the metabolism switches from burning glucose secreted by the liver to then transporting triglycerides from adipose tissue. These triglycerides are then converted into ketone bodies by the liver and can power the cells of the body in place of the depleted glucose. As well as fuel, ketone bodies are also signalling molecules: they inform each cell that there is a period of nutritional stress and the body is forced to become more efficient. Antioxidant defences are strengthened. DNA is repaired. Old proteins are digested and recycled. Life is made more resilient in food's absence.

For a polar bear or an elephant seal to endure months without eating is one thing. To then immediately feast is an evolutionary marvel that we still don't understand. In humans, such an immediate return to full meals after starvation leads to a so-called refeeding syndrome. A potentially lethal condition, returning to a normal intake of food after days or weeks without has long been known to lead to death. Once, after famine or the release of prisoners of war, the return of sustenance was a mysterious killer. Since the end of the Second World War (and the release of both prisoners of war from Japanese internment camps and concentration camp victims), however, the biochemical reasons for these deaths have been understood in detail. As a syndrome, there are many complications that can ultimately lead to death. But the main one is a lack of phosphorus, a mineral that is vital to the integrity of cell membranes, the provision of energy within the cell and the transportation of oxygen around the body.[11] Already depleted in bodies that aren't consuming this mineral in their diet, the return of food can use up the last of the meagre stocks and cause death through, among other things, cardiac arrest. Even today, refeeding syndrome is a common but often overlooked part of medical treatment. For people with depression, alcoholism or anorexia, not eating for five days can set the stage for problems once eating is resumed.

Fat sustains and supercharges life through hard times. In recent decades, however, this energy-rich resource has become mired by the unfortunate nature of some of our most common pollutants: poly-fluoro-alkyl substances (or PFAS) that are known as 'forever chemicals' – they are all lipophilic, 'fat-loving'. Since polar bears are at the top of the food chain, they accumulate the pollutants from everything below them. From the photosynthetic algae that feed the crustaceans that feed the seals, all of the forever chemicals and mercury are passed along the chain of predators and prey. In short, Andrew Derocher says, 'polar bears are highly polluted'. It isn't unusual to find *hundreds* of different industrial pollutants in a bear's tissue. One recent study led by organic chemists couldn't even identify many of the chemicals that had made their way into their samples. Released into the environment from industrial manufacturing, PFAS can also be found closer to home. Any material with a waterproof coating – whether it be raincoats, boots or tents – slowly shed their protective covering. That covering is made of stain-resisting, water-shedding PFAS. Adding to the problem, wind patterns and ocean currents have a tendency to push these chemicals towards the poles. The choices we make when we buy something with a click of a button, a tap of a screen, can have far-reaching impacts on the most remote places on Earth.

Whether these chemicals are actually harming the bears is another question. A study from Heli Routti, Sabrina Tartu and their colleagues found that some forever chemicals lowered the rate at which female polar bears could produce (and therefore store) fat. They also found evidence that they affected the thyroid hormones that are involved in 'synthesis, mobilization and degradation of lipids'. Studying 119 bears on the Norwegian island of Svalbard, Routti and her colleagues were working when sea ice was only 12.5 per cent compared to previous years. Most of Svalbard was ice free. Since females are fasting for four to eight months of the year *and* nursing their young, many were in poor body condition and their blood samples showed signs of 'nutritional stress'.

Female polar bears lose around a kilogram in weight every day they don't eat. Over an eight-month fast, 240 kilograms have disappeared – fat burned to keep their body warm and functional, their milk flowing into their young. Like steam released from a simmering stew, these fat cells are lost but the forever chemicals they contain are concentrated in a female's bloodstream. By the time her cubs are born, tiny fluff balls less than a kilogram in weight, her milk is sourcing its fat directly from her polluted tissues. 'The cubs are now getting this really high dose of pollution from their mother,' says Derocher. Adding to the work from Routti, human studies show that such pollution can impact learning ability later in life. 'If you don't learn properly in the Arctic,' Derocher says, 'you die.'

And yet, every polar bear researcher I have spoken to is shocked – and buoyed – by the perseverance of the populations they study. In Churchill in Canada, in eastern

Greenland, in Svalbard, the same story: they are heavily polluted, the ice they hunt on is forming later in the year and disappearing earlier, and yet they remain. 'They're smart,' says Kristin Laidre, a polar scientist at the University of Washington. 'Polar bears don't just kind of stand there on the beach… They're gonna try to survive.' Seabird eggs and whale carcasses can be a lifeline to see them through the extended periods without ice. Some polar bears on Svalbard have started to hunt reindeer when the ice disappears, forcing their prey into water where the polar bears are most adept. In Greenland, where Laidre studies the populations (with the aid of a helicopter), an isolated population of polar bears can extend their short season by hunting seals that haul themselves out onto glacier ice. But in places where ice no longer supports their prey, a polar bear's niche can shift northward – to places once too inhospitable for even *U. maritimus*.

The Arctic is formed of two types of ice: annual and multi-year. The former is where bears hunt, the seasonal layer of ice that encrusts the ocean's surface. Multi-year ice is much thicker because, as the name signifies, it has formed over many years. Like geological strata that lie under our feet, this ice is formed of layers and can be metres thick. Crucial for cooling the planet and keeping sea levels down, 'the multi-year ice is not a productive ecosystem', Laidre says. 'It's too thick, the sunlight can't penetrate. But as that ice thins, it does get better. And you do actually end up with more species and more productivity.' Seals can maintain their breathing holes, sunlight can promote the growth of algae and other plankton that float at the base of the food chain. As the Arctic melts, new ecosystems are taking their first breaths; an ever-changing platform of seals, their pups, and a hungry, hypertrophic species of white bear.

This is a transient benefit, Laidre cautions. The productivity of the Arctic will move towards the North Pole, but there's only so much planet for these ecosystems to shift into. Once the multi-year ice becomes annual ice, the next step in the chain is annual liquid water – an ocean at the top of our planet. 'We can make projections about different scenarios, but they really vary depending on what model you use and what inputs you use about human actions and human emissions,' says Laidre. 'There's a lot of good evidence that in the next decade or two, we'll have what's called a summertime ice-free Arctic.'

What this northward shift represents is an opportunity to curb our carbon emissions, an hourglass of ice dripping into water. 'How fast that habitat goes depends on our actions, globally,' says Laidre, 'and how fast we warm the planet.' There's no doubt that current projections paint a bleak picture. The melting of the Arctic doesn't just change polar bear habitats but all the other life that has evolved to live in this seasonal world of ice. Where will seals raise their pups? Where do narwhals find protection from predatory killer whales? What happens to the culture of Inuit people whose livelihoods have depended on a regular annual ice cover? To be a polar scientist working today, Laidre tells me, is a lot like being an oncologist. 'If you focus on how depressing your job is every day and how sad the news you're delivering, it would be a terrible job,' she says. 'But you kind of have to remove yourself and be a little bit more objective about the facts.' Sure, the ice is melting. Some polar bears are starving. But there's still a chance of recovery; a large population of blubber-loving bears that can continue to hunt seals on a solid layer of annual ice.

'We have local bears in Svalbard that do not move away from the islands despite the different sea ice conditions,' Jon Aars, a scientist at the Norwegian Polar Institute, tells me. In these areas, the length of time between the ice melting in spring and reforming in autumn has shifted by two months over the last two decades. 'They are doing OK, their condition is not worse,' Aars says. 'They reproduce, they survive. For the future, I would guess that at some point, it will start to be more difficult to be a polar bear there because it's predicted that the loss of sea ice is going to continue. We do think there is a limit; they need to be able to have seals on the sea ice for part of the year to be able to survive.'

While the common poorwill hibernates when the food runs out, most birds prefer to use their exceptional ability to move from one place to another. With stiff, feathery wings, most birds can take to the skies and flap, soar and glide to new locations. Instead of waiting for food to return, poorwill-style, they follow its seasonal shift. Once thought to be homebodies, animals that rarely moved from one place to another when times get tough,[12] bird migration has now been revealed to be a global event, a constant whir of wings that is powered by fat cells. Sanderlings, a small white-breasted sandpiper no bigger than your garden blackbird, flies from the Arctic breeding grounds to winter in the southern hemisphere, a 10,000-kilometre migration that takes in several stopover sites for refuelling. Arctic terns fly from pole to pole every year, a journey of up to 80,000 kilometres.[13] The swallows that swoop through the skies of the British summer have travelled more than 9,500 kilometres from their wintering grounds in sub-Saharan Africa.

Different routes, varying distances, but all sharing a similar purpose. By migrating long distances, the good times can be extended; the inhospitable avoided altogether. Instead of enduring a seasonal famine or the freezing dark of a northern winter, migratory birds concentrate their hardship into a few days of extraordinary athleticism.

Nowhere is this more apparent than in the bar-tailed godwit, an ordinary-looking shorebird that flies non-stop from northern Alaska to New Zealand, a journey of over 11,000 kilometres across the Pacific Ocean. In 2007, one female known as E7 completed this journey in nine days. Without food or water or sleep, her metabolic rate was thought to be ten times higher than at rest and her wings were flapping for the entire trip. 'These non-stop flights,' one paper stated in 2009, 'establish new extremes for vertebrate performance.' The maintenance of such a high metabolic rate without food or rest for nine days, the authors added, 'is unprecedented in the current literature on animal energetics'.

It's almost impossible to comprehend the endurance of these birds. It might help to remember that the best marathon runner in the world is running non-stop, without food, for two hours. In 2022, a male bar-tailed godwit was recorded in flight for 237 hours (nearly ten days). To fuel such a migration every year, a bar-tailed godwit has to become incredibly fat, gorging on clams and worms buried in the Alaskan mudflats – in particular the

Yukon–Kuskokwim Delta, a haven of sand and mud that isn't just rich in invertebrates but also low in predators. Using their long, slightly upturned beaks, godwits prod the shallow water like the needle of a sewing machine, constantly moving up and down to consume enough calories over the Arctic summer. Whether protein or carbohydrate is digested, the excess calories are converted into fat, stored inside cells called adipocytes located around the body. 'The adipocyte is basically like a droplet of oil surrounded by a thin cell membrane and a nucleus,' says Chris Guglielmo, a researcher from the University of Western Ontario. 'And they can swell and shrink. So these adipocytes actually grow in size as the birds get fatter.' Primarily stored in an abdominal fat deposit and underneath the skin, fat cells also line the intestines of these birds. A godwit has both visceral and subcutaneous fat, in other words. By the end of the Arctic summer, the body mass of some birds is over 50 per cent fat. 'A godwit in August in Alaska is like a flying softball,' says Bob Gill, a retired Arctic field biologist who still publishes studies into the bar-tailed godwit's migration. 'It is so round, engorged in fat, this little tiny head sticking out, the breast and abdomen are super extended, and it's just a thick layer of fat under there.' He has noticed that birds struggle to take flight without the help of a favourable wind. In a 2018 paper, Guglielmo calls these birds 'obese super athletes'.

Fat and athletic endurance don't usually go hand in hand, and yet obesity is by far the best option for a migrating bird. Gram for gram, fat can release eight times the amount of energy than glucose. This is due to the chemical structure of fats. The type of fat that animals can use to power their cells are called fatty acids, and they are made up of long chains of carbon atoms and hydrogen atoms, each ending with a small acid group at the end. Only this last part, the acid, has a couple of oxygen atoms as part of its structure. A chemical formula might look like this: $C_{16}H_{32}O_2$. Now, compare this to glucose: $C_6H_{12}O_6$. The ratio between carbon atoms and oxygen atoms is much higher – 1:1 in glucose and 8:1 in this particular fatty acid. And this is important when these molecules are burned aerobically (in the presence of oxygen). Because glucose has a high proportion of oxygen in its chemical structure, a large part of its weight is composed of atoms that can't release any more energy. Instead, oxygen is combined with hydrogen to make water. But fatty acids are almost entirely made up of carbon and hydrogen atoms, the same as the hydrocarbon chains that power vehicles on our roads and planes through the sky. 'Fatty acids are more like gasoline,' Guglielmo says.

Fat is also water insoluble, meaning that, unlike proteins or carbohydrates, it can be stored in a dried state, thereby saving weight. 'Fat is the logical choice for storing a lot of energy for very little weight,' says Guglielmo, 'which, if you're a flying animal, is the most important thing to economize.'

But here's the catch. Unlike glucose, fatty acids can't simply be secreted into the blood and expected to reach the muscles that are powering flight. The insolubility of fats might make them lightweight but it also means that they can't diffuse through an animal that is 60 per cent water. To get around this not-insoluble-insoluble problem, migrating birds like bar-tailed godwits pump their bodies full of molecular guides that can latch onto fatty acids secreted by adipocytes and shepherd them to where they are needed – primarily the flight muscles and lungs. 'And that's where mammals fall behind,' says Guglielmo. 'We're terrible at the transport side [of fatty acid metabolism]. Even when we run marathons, we have to rely on glycogen and glucose... we don't rely on fat very well.'

By investing in these transport proteins just before their long-distance migrations, however, birds like bar-tailed godwits can quickly and easily tap into the richest sources of energy in the animal kingdom.

Even when they land in New Zealand, these birds don't immediately begin a post-workout feast. They need to sleep. While there is evidence that some birds can drift in and out of sleep during long-distance flights over the ocean, it is clear that the godwits still have to recharge. 'They arrive, tuck their head under their feathers on their back and just zonk out for the day,' says Gill. After ten days of non-stop flying and a day of sleep, only then do they start to feed in the austral summer. By March, they return to the northern hemisphere, stopping off in the Yellow Sea in China, before flying 5,000 kilometres back to the Arctic tundra to breed and feast once more.[14] While a common poorwill has found evolutionary success by not moving an inch during the harshest winters, a bar-tailed godwit might fly more than 500,000 kilometres over its lifetime, a distance that would easily get it to the moon, just to avoid them.[15] To escape the toughest times, at least for bar-tailed godwits, means pushing a body to near-complete exhaustion, an extreme means of survival that only birds – with their refined fat transportation systems – can accomplish.

PART TWO

Atoms in Motion

CHAPTER FOUR

SUPERCOOL ANIMALS

Freezing

For most of its history, Antarctica has been a warm place. Grouped together in the southern hemisphere, Antarctica, South America, Australia, Africa and India formed a generally warm southern supercontinent called Gondwana. Much of the world to the north was one giant ocean: Pantalassia. From 200 million years ago, powered by the inner turmoil of the planet, this continental kinship started to break apart, forming the landmasses we are familiar with today. Africa moved to the equator. India zoomed through the ocean and collided with the Asian subcontinent, a collision that formed the tallest mountain range in the modern world, the Himalayas. Australia and Tasmania became marooned in Oceania. Stretching out a long tendril of earth, the Antarctic Peninsula held onto the southern tip of South America for much of this story of separation. Around 30 million years ago, however, a time when mammals were growing into a wonderful array of sizes and shapes – giant sloths and car-sized armadillos, to name two – this terrestrial umbilical cord was severed. The Atlantic and Pacific Oceans flooded into the gap, collided and birthed a new ocean that was nothing like its parents. Surrounded by the swirl of the Southern Ocean, Antarctica slowly drifted into a deep freeze.

With no landmass to stop its flow, the Southern Ocean swirled around this freshly isolated continent unimpeded. No longer connected to the tropical influence to the north, Antarctica was bathed only in its own roiling circumpolar current, a barrier to almost all outsiders. Warmth couldn't penetrate it. Few animals could cross it. For the most powerful swimmers such as migrating whales, this invisible line would be felt as a sudden drop in temperature, a significant three or four degrees within a few kilometres of ocean. This was the moment they had crossed into another world.

Since its separation from South America, Antarctica has been steadily cooling. The first ice sheets formed quite quickly, geologically speaking, perhaps around 30 million years ago. But it wasn't until 14 million years ago that it became a continent of thick ice sheets and coastal waters filled with pack ice. With steep mountains separating the much smaller western Antarctic from the endless ice plains of the east, it is also a windy place. Katabatic winds develop from the very cold and dense air that is pulled from the mountain peaks by gravity. Picking up ice crystals and snow as they move downwards, they can blow at an abrasive 130 kilometres per hour for weeks on end. For eight months of each year, it is either totally dark or endlessly light.

And then there's the temperature, unlike anywhere else on Earth. On 21 July 1983, Russian scientists working in the austral winter of eastern Antarctica recorded a temperature of −89°C, the coldest place on Earth ever recorded at ground level. Since then, satellite-based recordings have logged temperatures of −98°C.

This is all to say that Antarctica is an isolated and hostile place. Imagine the surprise on Third Officer Reginald Skelton's face, then, when he had spent weeks battling over the treacherous Southern Ocean aboard the RRS *Discovery* in 1902, surmounting some of the highest waves in the world, and spotted a group of birds not only living on the pack ice along the Antarctic coast but raising their newly hatched young on it. With their fluffy chicks huddled underneath a flap of skin that folds over their feet, emperor penguins must have seemed like a joke made by Mother Earth. *You think your trip was difficult? These birds happily swim among those icebergs you passed; their first sensation of solid ground is ice.*

Later explorers would find a second species of penguin, the Adélie, named after Adélie Land where explorer Jules

Dumont d'Urville's team collected the first specimens of this species. (Adélie Land was, in turn, named after d'Urville's wife, Adéle.) Smaller than the emperor penguins and without the flashy orange cheeks, Adélie penguins have a thick bristly tail that helps them gain purchase on their preferred rocky outcrops, jet black feathers all over their shoulders, wings, head and most of their beak, and a pale blue ring around their otherwise black eyes. They fit the cute stereotype for a penguin. As William Clayton wrote for the Royal Society of London in 1776, the Adélie has, 'at first sight, the look of a child, waddling along with a bib and apron on'. (Although childlike in appearance, this exterior didn't stop explorers from eating their eggs and flesh. As Clayton added in the same paragraph, 'All their eggs are good nourishing food, and a great refreshment to the seamen; but the flesh of these animals is coarse, fishy, and wholly unfit to eat.')

But their childlike appearance is deceptive. With deepest respect to Madame Adéle's legacy, these birds are not the embodiments of childlike innocence that early scientists pushed into the public eye. Living in a harsh place can imbue a ruthless edge to the circle of life. With a short season in which to breed on the sea ice, a male Adélie penguin's body is pumped full of hormones in order to maximize breeding success. And this makes them, as the author and broadcaster Lucy Cooke put it, 'have sex with anything that moves, as well as some things that don't, including dead penguins'. Rape, necrophilia and sex with pebbles: these are the realities that explorers had to keep secret on their return to Victorian Britain. When books were written on the Adélie penguin, Cooke notes, the less than savoury accounts were left out, to be published separately – and discreetly.

Blimp-shaped seals, penguins huddling together against a blizzard's fury, pods of killer whales breathing into the subzero air: these are images of Antarctica that exist in our minds. But the oft-ignored wonders of this continent lie in its depths, below the sea ice that grows and thins with the seasons. The Southern Ocean contains an estimated 20,000 species of sea-dwelling animals, a similar level of biodiversity to every other ocean ecosystem except the kaleidoscopic coral reefs of the tropics. 'But we've only got names for 8,000 of them,' says Melody Clark, a molecular biologist at the British Antarctic Survey. Out of these 8,000, she adds, the life cycles and ecological relationships with other animals are only known for a 'small handful of the most common species that you find near research stations. So there are huge swathes of the unknown.'

Each species has been carved by the history of ice. Over millions of years of temperature fluctuations, the ice around Antarctica has grown and shrunk, extended and retracted over the seabed. This process of separation and reintroduction is one of the basic ingredients for evolutionary change. Animals that have been separated can follow different evolutionary paths. For this, the ice of Antarctica has been called an 'evolutionary pump', a generator of species diversity. Add to this the fact that cold water can hold more oxygen, and life in these waters has been able to power some truly exceptional forms. Sea anemones the size of buckets. Starfish with

40 ray-like arms that grow to the diameter of mountain bike wheels. Sea spiders, no relation to terrestrial arachnids, with bodies so small that their reproductive organs and digestive tract reach into their legs. And then there are the fish, including 16 species of notothenioid 'icefish' that live in −1.9°C degrees and, in the words of Lloyd Peck at Cambridge University, 'do life differently'. Out of 40,000 species of vertebrate – every fish, salamander, frog, lizard, snake, bird and mammal – only these 16 species lack haemoglobin, the oxygen-transporting protein that keeps the fires of an animal's metabolism burning.

The waters surrounding Antarctica can also be defined by their exceptions. Unlike in the Arctic, there are no sharks, only a few skates, one species of codfish, and no crabs or lobsters. For the latter, it is the cold itself that made these waters uninhabitable. Reptant decapods, the group to which crabs and lobsters belong, have high levels of magnesium in their blood. In freezing conditions, this metallic element has narcotic-like effects, transforming a careful crustacean into a careless one. One species of crab that lives in the Arctic spends several months rolling around on the seabed, seemingly sedated and easy pickings for any shell-crushing predators. It breeds in those warmer months when it can function. In the Antarctic, with its perpetually cold temperatures, there would be no season in which the necessities of life could be pursued.

In the absence of these common groups of animals, Antarctica represented a vacant niche to expand into. The reduced competition, when combined with Antarctica's geographical isolation over millions of years, has created a unique assortment of marine animals. The term used for a species that can only be found in one place, whether a country or a continent, is 'endemic', and Antarctica's waters are full of them. Half of all sponges. Three-quarters of molluscs. Over 90 per cent of sea spider species are endemic. For the most abundant fish group, the notothenioids that include the 16 species of bloodless fish, 97 per cent of the 139 known species are only found in the Southern Ocean. Their dominance in the Southern Ocean is perhaps better encapsulated by their collective weight: 90 per cent of all the fishy biomass in the Southern Ocean is made up from notothenioids. As one Antarctic researcher puts it, Antarctica has 'the world's most distinctive marine biota'. Coral reefs might attract the zoologically uninitiated, but the deep waters surrounding the South Pole are for the connoisseur.

In the nineteenth century British whalers called them 'icefish', a name that derives from their clear, water-like blood and the icy places they are found near to the coast of Antarctica. With large heads, long snouts and rows of sharp teeth, they have also been called crocodile icefish. So fearsome is their visage, their Latin names often refer to dragons. One species belongs to the genus *Dacodraco*, a portmanteau of 'one that bites' and 'dragon'. Crocodiles and dragons may give the impression of a large animal but most species of icefish are about the size of your foot. The largest can grow to half a metre. Still, for the smaller fish that make up their prey, they are literally cold-blooded killers.

Like crocodiles, icefish are ambush predators, conserving their energy until the moment they need to strike. *Dacodraco* can eat an Antarctic whiting, a shoaling fish half its length, in one sitting. Its jaws are slightly curved, only meeting at the front and rear, creating a perfect set of pincers to grip a relatively large and strong prey. Its row of sharp canine teeth impale from all sides, ensuring a quick death and little struggle. Everything in these subzero waters necessitates economy.

Even ambush predators require a burst of energy to catch their prey. Acute senses, an active mind and powerful muscles need fuel – the combustion of oxygen. So how do icefish hunt without haemoglobin? This question has fascinated biologists since these fish were first discovered. One answer is that by living in cold waters, a place where dissolved oxygen is higher than in more temperate regions, they don't need to actively transport oxygen to their cells. Without red blood cells, and the haemoglobin they contain, it is a more leisurely process, governed by diffusion. It's as if the fish have become part of their environment, just another place where the strong currents of the Southern Ocean swirl in and out. The second answer is that they have compensated by evolving relatively large hearts, wide blood vessels, and spend a lot of their energy powering this big cardiovascular system. While skipjack tuna, an athletic sprinter of the open seas, may expend 2 per cent of its energy budget on its beating heart, icefish of the Southern Ocean use over 20 per cent.[1] Even when compared to their notothenioid relatives that produce red blood cells and haemoglobin, their hearts are still working twice as hard to pump blood around their body. While icefish lazily float under pack ice or rest on the silty seabed, their hearts are working overtime.

This begs the question of why they lost haemoglobin in the first place. The most popular theory for a long time was that this watery blood was less viscous and easier to pump around the body. In freezing cold temperatures, this may have reduced the amount of effort needed to circulate oxygen. But knowing how hard their cardiovascular systems are working seems to cast some doubt on this theory. More recently, a few scientists have proposed the idea that there is no advantage to this loss of haemoglobin. It is an example of a 'disaptation', a trait that has made these fish less adapted to their environment compared to their ancestors. It is survival of the less fit.

Disaptations seem to run counter to the general idea of evolution by natural selection. But it can be explained when seen within the concept of an ecological niche. Icefish evolved in a harsh environment where few other fish can survive. With less competition, it is a 'relaxed niche'. With the formation of the Southern Ocean it was as if the ancestors of the icefish were adrift on an enormous island of possibility. In this relaxed state of evolution, weird things can happen, like red blood turning clear. It may have been slightly disadvantageous, but not enough to make a difference to a fish's survival.

'The jury is still out on that part,' says Iliana Bista, a geneticist at the Senckenberg Research Institute in Frankfurt, Germany, who sequenced some of the first genomes of icefish in 2023. 'I wouldn't rule it out. But I couldn't support it either.'

Although it lacks red blood cells and haemoglobin, icefish blood contains other proteins that have been crucial – beyond doubt – to their survival in frozen oceans. These are known as antifreeze proteins, and they allow many species of notothenioids (not just icefish) to maintain a liquid state when the world around them starts to freeze. These 'antifreeze glycoproteins' are ten times as effective as the ethylene glycol used in windscreen de-icer. The mechanism of this biological antifreeze has only recently been studied in detail, and it seems that these proteins bind to ice crystals that have already formed, bending them into a position that prevents further growth. The amount of energy required to make a speck of ice into a small crystal is increased by this process. Without these proteins, fish that inhale the ice crystals that continually float through the shallow waters of the Antarctic coastline would be shredded from the inside out. The seabed would be devoid of fish and, in turn, the seals and whales that prey on them. Without this 'biochemical cunning', as Arthur DeVries, an icefish biologist, calls it, Antarctica would be a very different place.

With their ability to live in subzero temperatures, icefish didn't just survive along the coastline of Antarctica, they thrived. While fishing with trawl nets and scuba diving can provide glimpses into the waters near the sea ice, the latest discoveries into their lives have come from remotely operated vehicles (ROVs). While a few of these submersibles can be controlled from the deck of a ship much like a remote-controlled car or drone, the most cost-effective approach is often to tow a mass of cameras and sensors behind a research vessel. This was what Autun Purser and his colleagues did as they explored parts of the Weddell Sea on board the German research vessel *Polarstern*. In the early months of 2021, just before the winter freeze encased their expedition in ice, they dropped a thick metal cage into the freezing waters. Weighing one tonne, the unit – known as OFOBS[2] – isn't pretty. It lacks the bright colours and torpedo-like shape of other submersibles. It's more like a lump of scaffolding or a few girders welded together. But its heavy-duty appearance has a purpose. Being towed behind the ship, it is at risk of colliding with the rough seabed or unseen rocky outcrops. Attached to this skeletal frame are a set of very expensive cameras, one for stills and another for video, sensors for temperature, water movement and depth, as well as a few acoustic cameras that use sonar to scan the seabed up to 50 metres in every direction.

Moving at a leisurely 0.5 knots (around a kilometre per hour), Purser and his team watched the feed on their laptops. As the OFOBS floated a metre or so above the Filchner Trough, 500 metres below the surface, they were met with a strange sight. Circular nest after circular nest were illuminated under the OFOBS lights, each with an icy blue fish curled around its contours or floating above in the gentle current. They had just discovered a mass breeding colony of *Neopagetopsis ionah*, Jonah's icefish. Over the next three surveys, each lasting for an exhausting 12 hours, they estimated that there were 60 million nests. Even increasing the speed of the ship, and therefore the amount of distance covered by OFOBS, they couldn't observe the entire breeding colony. In an otherwise featureless plain, the scale of each nest and fish is difficult to capture from a single image or video. But each nest is the size and depth of a large paella pan or the crash cymbal on a drum kit. *N. ionah* can grow to half a metre long. And each female can lay upwards of 2,000 eggs.

(Although some species of icefish can lay 15,000 eggs, the average is 1,500 or so.) With the same icy blue tint as the parent, the large eggs (four to five millimetres in diameter) are attached to a coarse layer of gravel at the centre of the bowl-shaped nest.

This wasn't the first time that scientists had seen the nesting behaviour of Antarctic fish. In the 1970s, another species (a red-blooded fish) was seen guarding its nest with regular displays of its sharp gill covers and pectoral fins. But these shallow-water species built their nests next to large rocks and rarely grouped together. The breeding site of *N. ionah*, however, was more similar to birds that breed on the rocky escarpments above, each nest almost in contact with its neighbours. Together, they create a draughts-on-a-draughts-board pattern on the seabed. The reason for this dense aggregation, as with birds such as gannets or Adélie penguins, is to avoid – or reduce – predation. For both, these threats come from above. Combining data from previous studies to their research, Purser and his colleagues found that satellite trackers attached to Weddell seals captured these deep-diving mammals at the same sites where they had observed the icefish colonies.

These 'mass colonies' of half-metre-long fish show how little we know of the waters that are out of sight. For decades, trawling has been a primary method of surveying the ocean's depths, but now images from submersibles like OFOBS show how important it is to understand the deep sea before we disturb it. Other notothenioids are known to attach their eggs to glass sponges, one of the main forms of animal life attached to the seabed around Antarctica.[3] Whether it's in nests or sponges, the wonders of icefish extend from their clear blood, their antifreeze molecules, all the way back to the moment they were a fertilized egg encircled by a protective parent.

There are more familiar ways to protect cells and organs from the damaging effects of ice formation. The wood frog of North America, for example, uses glucose, a simple sugar molecule we extract from longer starch molecules in our diet, to survive winter temperatures that can drop to −18°C, the same temperature as a household freezer. Known as a cryoprotectant, glucose holds a cell together as it freezes, preventing it from breaking open as ice formation sucks all the water out of its innards. Freezing, at its most basic level, is a type of dehydration stress. Whether water is lost to ice formation or to the environment results in the same thing: a cell shrinks, cell membranes start to stick together, proteins begin to lose their all-important shape. This is what causes frostbite: the cells of fingers or toes lose so much water to the surrounding ice that they rupture.

A wood frog needn't worry about frostbite. Constructing a hibernation chamber under leaf litter near their natal pond, these frogs pump so much glucose into their cells that they don't split open as water is forced into the ice crystals forming in between cells. As the first layer of ice coats their skin, the frogs start releasing glucose from their liver, pumping it through their bloodstream to every organ and cell in the body. Over many freezing nights and thawing days, this sequence continues. By the time winter descends, daytime temperatures no longer provide any warmth for thawing. The Alaskan winter and a hardy amphibian have combined to create its most wondrous product: an intricate block of ice that follows the contours of muscle fibres, the spinal fluid and, finally, surrounds the heart of the wood frog. In total, over 65 per cent of the frog's water has frozen solid.

Until the spring thaw, each organ system is kept on life support, disconnected from each other in both communication and function. The heart, brain, lungs and digestive system are no longer part of a larger whole. The frog, it can be said, has ceased to exist. 'It's not an organism anymore,' says Don Larson who studies wood frogs and their parasites in Fairbanks, Alaska. 'It's *parts* of an organism that can come back and revive.' He adds that there's a little bit of Frankenstein's monster in these animals, although their organs aren't coming from separate individuals. While damage is inevitable, it can be repaired upon thawing, a process that, oddly, begins from the inside out. The heart is the last organ to freeze and the first to be freed from its encasement of ice. Blood then begins to flow back through the body. Breathing returns. A leg muscle contracts.

As we learned in the chapter on anoxia, glucose is the basic metabolic fuel of all cells. It is already a staple of an animal's daily function. Freeze-tolerant frogs simply produce an excessive amount of it. Samples taken from wood frogs have found nine grams of sugar per 100 millilitres of blood, an extreme level of hyperglycaemia. A human body might only contain four grams *in total*. 'If you were to take a sample of wood frog's blood and inject it into yourself,'

Larson says, 'you would kill yourself very quickly. It has incredibly high levels of sugar.'

The fact that wood frogs can freeze solid for up to seven – seven! – months every winter has enabled them to live at latitudes that no other amphibian in North America can tolerate. But they also live as far south as Ohio, a place where a freeze might last for a couple of days and never falls below −10°C. As a species, the wood frog is adaptable, an amphibian that can live in water and on land, in a frozen state and a liquid state. As with painted turtles and magnets, they are attracted to opposites.

As with surviving under an ice-covered pond, the advantage of freezing solid over winter is to have first dibs on spring. 'Wood frogs have no competition in these northern regions from other frogs, and less predation from snakes and things that might eat them,' says Brian Barnes, an Arctic biologist based in Fairbanks. 'And so having evolved this ability to tolerate freezing opened up the northern hemisphere to them to invade and prosper. In Canada, they've reached the Arctic Ocean; they're right at the northernmost lands. In Alaska, the Brooks Range, the northern mountains, curve around and have blocked the dispersal of frogs into northern Alaska proper, so far.' It takes a mountain range to stop the wood frog from invading and prospering. And even then, there are unconfirmed reports of frogs in these regions, Barnes adds.

A key to their success might be their spontaneity. While they build up their glucose reserves over the summer, they don't actually begin to release it into the bloodstream until they are sure they need to use it. 'Other animals make their antifreeze proteins and their cryoprotectants in preparation,'

Larson says. 'The wood frog does it responsively, reactively, and that's unique.' It's as if they are still holding on for a warmer winter, hoping that they might not have to spend the winter as an inanimate block of ice. But the initial touch of frost triggers adrenaline, the same neurotransmitter that our nervous systems use to initiate a flight or fight response, and the process of glycolysis – releasing glucose from its storage inside the liver – begins. Hidden underneath a pile of leaves and a pack of snow, the frogs are unable to fly or fight. They acquiesce. As the world around them freezes, they freeze with it.

If you're a biological system of delicate cells and complex organs, ice is generally a bad thing. Solid water is a crystal that grows like cancer and shreds tissue like shards of glass. The wood frog controls its formation to control its damage. Icefish try to stop its spread entirely. Our own relationship with ice is also often coloured with tragedy; car crashes, research vessels crushed by pack ice, the *Titanic*. But there are organisms that actively promote the formation of ice for their own benefit. Ice, to them, isn't harmful but necessary.

Originally discovered on agricultural plants such as apple trees, olives and wild cherry, *Pseudomonas syringae* is a type of bacterium that nudges water in the opposite direction to an icefish's antifreeze proteins. Its membrane is peppered with a specific arrangement of proteins that hold water molecules in the arrangement of an ice crystal. They are a type of 'ice nucleator', a central node on which water can construct a crystal. Other types of nucleators include dust from eroded rock and desert sand, fungal spores and pollen. Such 'impurities' are crucial for ice to form at subzero temperatures. Without them, a sample of pure water wouldn't freeze until it reached −38°C. This absence of freezing below zero degrees is known as supercooling, a thermal range where water remains in the liquid phase all the while having the potential to snap into a solid.

Supercooled water might sound like an unnatural product of chemistry labs and their sterilized glassware. But it can be found over 60 per cent of the Earth's surface at any one time. It's likely that there's some above you right now. Water in clouds is supercooled.[4] Held in a mixture of liquid and vapour, the water that makes up clouds can be well below −20°C.

Ice nucleators can change this. *P. syringae*, for example, has been sampled from clouds thousands of metres above the ground, a population of microbes that are carried into the skies on updraughts – either alone or hitchhiking on fungal spores – and can live and reproduce in the troposphere. In a matter of days, they can be transported hundreds, if not thousands, of kilometres. When found in sufficient numbers, these bacteria grab hold of water in its liquid or gaseous state, position each molecule in the arrangement of an ice crystal, and kickstart the transition into a solid. 'They have a protein on the outer membrane,' says Cindy Morris, a microbiologist who studies plant pathogens such as *P. syringae*. 'It's this beta pleated sheet that just so happens to have places where water molecules can bind in a way that it makes a configuration of ice.' As it grows, this bacteria-induced ice crystal gains weight and eventually falls from its supercooled cloud. It precipitates. Snow, sleet, hail or rain: the nucleator is carried back to Earth like a space shuttle returning astronauts to terra firma.

'All micro-organisms can go up and ride the clouds,' Morris adds. It's getting back down that *P. syringae* has mastered.

Not every drop of rain contains an ice-loving bacterium. In the tropics and subtropics, rain forms from the movement of water molecules in cumulonimbus clouds that tower into the stratosphere. The constant jostling ensures that smaller drops bump into each other, forming larger droplets that can fall to the ground. But in more temperate and polar regions, in clouds that contain water in both liquid and solid ice – known as mixed phase clouds – ice nucleators are often key to the formation of snow and rain. A study that sampled water sources in France, Italy, Austria and the United States found *P. syringae* everywhere, a natural component of the hydrological cycle. 'In rain, in snow-melt water and in lakes and streams, *P. syringae* was present at concentrations generally between hundreds to several thousands [of] bacteria per litre,' Morris and her colleagues wrote in 2014. Once thought to be only a pathogen of crops, a shadow that hovered above agricultural fields like an invisible plague, this species of bacteria is found in so-called pristine areas with no connection to croplands. Morris and her colleagues argue that *P. syringae* has evolved to spread from a plant's surface, enter the atmosphere and influence the weather. By controlling the freezing point of water, these ubiquitous microbes increase the likelihood of rain. After travelling for hundreds of kilometres in a mixture of water and ice, they fall from the skies and land on a new plant host in a new place.

This theory of the interrelationship between bacteria and the weather is called bioprecipitation. To push the phenomenon to its metaphorical limit, it's as if this pathogenic microbe infects clouds with rain. In a white and fluffy stratocumulus cloud that hovers between one and two kilometres above the ground, the proteins on the surface of these bacteria manipulate water to their will, ensuring that they don't remain in the skies – a low-resource, freezing-cold setting – for too long. As with a sneeze or a cough, the bacteria is ejected from its ephemeral host in a shower of precipitation that ensures its spread.

To see weather as an abiotic phenomenon, therefore, completely ignores the biological heart of this planet. Rain can also be biotic, a fall of water that life encourages in order to create more life.

Far below the wisps of cloud that transport pollen, fungal spores and microbes to new territories, an Arctic ground squirrel begins to dig her burrow. In the foothills of the Brooks Range of Alaska, she can't excavate far into the soil. Just a metre below the surface, the ground remains frozen solid. As she reaches this boundary of permafrost, the squirrel creates a spherical chamber at the end of her burrow, curls up into a ball with her fluffy tail over her head, and begins to ramp down her metabolism. As metabolism releases heat, this squirrel can reduce her thermostat to a point when she is −3°C – a subzero squirrel. Textbooks often state that hibernators such as Arctic ground squirrels become cold as the world around them freezes, and that this slows down their metabolism. But it's the other way around: these tiny mammals decrease their metabolism, which then allows their bodies to cool.

Unlike the wood frog, however, this squirrel doesn't freeze. Like the water in the clouds above, she supercools. Before hibernation, 'they cleanse their blood of would-be nucleators prior to entering the hibernation state in the Arctic,' says Barnes, the Arctic researcher in Fairbanks. By fasting in the weeks before the harsh winter, these squirrels reduce those bits and pieces to which ice crystals grow upon. This trick has made ground squirrels literally the coolest mammals on Earth.

Many species of insects that experience freezing conditions for part of their life cycle do the same. But they can supercool even further.[5] The current record holder for an insect that keeps its body in a liquid state in subzero conditions is the red flat bark beetle of Alaska. Its larvae, only a centimetre long and huddled together underneath the bark of a fallen poplar, can fend off freezing at −150°C in the lab. The coldest temperature they may ever experience in the wild might be between −60°C and −70°C. At these near-Antarctic levels of cold, these beetle larvae become impervious to freezing, no matter how much colder they get. As with tardigrades and other anhydrobiotes we met in Chapter 1, their cells vitrify into a form of biological glass at −60°C or so, a state that has made any other transformation into another solid (such as an ice crystal) almost impossible. By utilizing anti-freeze proteins and cryoprotectants (like glucose), and by dehydrating its body, removing ice nucleators in its gut and then vitrifying its cells into glass, this beetle has been called 'unfreezable' by the scientists who study it.

Arctic ground squirrels are the only mammal known to supercool. For the dark and cold six months of an Arctic winter, in other words, they are more akin to freeze-avoiding insects than the bears that they share their environment with. The classic example of a hibernator, the black bear, for instance, only cools its body to three to five degrees below its summer norm of 37°C. This is because they only reduce their metabolism during hibernation to as low as one-third of normal summer rates. Their large body size also holds onto more heat, a product of its low surface area (where heat is lost) to volume (where heat is produced) ratio. A ground squirrel, however, can reduce its metabolism by up to 99 per cent compared to its summer and spring norm.

A ground squirrel weighing the same as a bag of flour and a bear that weighs more than a professional rugby player hibernate in very different ways. One switches off its metabolism almost entirely while the other maintains a level of activity. This is also reflected in their responsiveness to disturbance. A black bear will open her eyes if she hears noises. In their artificial dens at the University of Alaska Fairbanks, for example, Barnes has watched as bears become more active during New Year's Eve when fireworks are exploding overhead. 'She startles, lifts her head and blinks in the dark,' he says. 'Then she just goes right back into hibernation and isn't disturbed by that.' A brown bear, however, is likely to emerge from hibernation completely and investigate what was causing the disturbance. 'Once every winter or two someone will be killed by a grizzly bear because they've tramped over its den and it's emerged and attacked them,' Barnes says. 'A black bear won't do that.'

Neither will an Arctic ground squirrel. For one, they are herbivores the size of a guinea pig. And second, their state of torpor is so deep that they remain blissfully unaware of anything that goes on around them. You could dig up a ground squirrel and it would remain curled up in a ball, its fluffy tail still tucked over its head for added insulation. But there are times when even a ground squirrel has to return to consciousness. Every three weeks or so, its body begins to warm up. Over a few hours a supercool −3°C becomes a 37°C endothermic norm. But you wouldn't notice any difference in the squirrel unless you had a thermometer. It remains deathly immobile, still tucked up in its pose of torpor. For 12 hours, the body temperature is maintained at this relatively extreme level of warmth (when compared to the surrounding winter). Studying this species for over 25 years, Barnes thinks that these bouts of warmth are proof that a hibernating ground squirrel has to emerge from hibernation to *sleep*. While hibernating, these mammals might look like they are sleeping but they are actually physically incapable of sleep. 'Their brains are too cold to sleep,' Barnes says. 'The synapses cannot fire to produce the waveforms of sleep, but they still accumulate a need for sleep, albeit slowly. And instead of having to sleep every morning for hours like we do and they do during the summer, they can put off sleep for three weeks. But then the pressure for sleep is so high. They warm up, sleep it off, and then go back into torpor.'

Currently, this is still a hypothesis, one that Barnes admits has its critics. Why we sleep is still a mystery in itself, never mind why an Arctic ground squirrel would warm itself up every three weeks. But in recent years, our perception of sleep has shifted from a state of relative brain inactivity to one of energetic maintenance. While the neurons of the brain do indeed reduce their activity during sleep, another type of brain cell – glial cells – ramp up activity. If neurons are the messengers of the brain, glial cells are the cleaners. All the metabolic by-products of the previous day are swept up and shuttled into a specialized network of fluid-filled vessels known as the glymphatic system. During the day, this system is relatively inactive, the vessels constricted, closed, like the capillaries of our skin during cold weather. When we sleep, however, the glymphatic system opens and begins to accept the waste from its roving workforce of glial cells.

How does this relate to a hibernating ground squirrel? Barnes explains that the freezing temperatures that their bodies experience every winter have an effect on this glymphatic system. As with a car's tyres during the winter, cold temperatures lead to a drop in pressure. And for a system that cleanses the brain of its waste products, pressure is essential. At −3°C, the glymphatic system of an Arctic ground squirrel can't flow with enough force to flush. While the waste products are also reduced in their production at these lower temperatures, they still accumulate. Instead of sleeping once a day, a hibernating Arctic ground squirrel has to sleep every three weeks to clear this trickle of brain excrement. 'The problem with getting through the winter is not how to survive not eating or drinking,' Barnes says. 'That's not a problem. It's not a problem of being cold, even to the degree of supercooling as Arctic ground squirrels do... The problem is getting enough sleep.'

Over half of the energy reserves that an Arctic ground squirrel uses during hibernation are dedicated to these periodic rewarming episodes. Whether it's sleeping or not, they are undeniably a significant part of its winter survival.

Frogs that can protect their vital organs and mammals that can supercool for three weeks might seem like phenomena that we can learn from for long-term tissue preservation. Indeed, some of the same molecules that have been discovered in these animals have been utilized in the field of cryopreservation.[6] The first sperm, for example, were frozen with glycerol in the 1950s, a type of alcohol that animals such as frogs and insects routinely use as a cryoprotectant. Similarly, synthetic molecules such as DMSO and ethylene glycol are cheaper to produce and more easily absorbed by cells than sugars like glucose. But even with these advantages, our attempts to preserve anything larger than a single cell – such as sperm, eggs or stem cells – are still in their infancy.

We can reconstruct the DNA of a horse that died 800,000 years ago and was preserved in permafrost, but we can't keep a human organ alive for more than a few hours. Kidneys, livers, hearts: they are too large and too complex to be protected from the ravages of ice formation. After three to 12 hours, a transplant will not work, the organ has lost its ability to function in another body. For organ transplants, supply is not simply ensuring there are enough donors. It is the rush to remove and reinsert organs before they degrade. In 2010, the World Health Organization found that only 10 per cent of the need for organ transplantation was being met. In the US, more than 70 per cent of all donor organs are discarded each year. As a 2021 review entitled 'Winter is coming: the future of cryopreservation' states, 'The lack of organs for transplantation constitutes a major medical challenge, stemming largely from the inability to preserve donated organs until a suitable recipient is found.' To increase the length of time an organ can be kept on ice, even by just a few hours, would be transformative.

Supercooling might hold the key. Just as an Arctic insect can hold off ice formation below −30°C, can we slow the biochemical slide into decay and prevent the formation of ice? A study published in *Nature* in 2019 suggests so. After supercooling rat livers, researchers from Harvard Medical School tested their methods on livers that had been rejected by human recipients and were to be thrown away. While the rat livers were supercooled to −6°C and transplanted after four days,[7] a human liver is a different beast altogether. It is 300 times[8] the size of a rat's liver, and size makes the possibility of ice nucleation more likely, as well as reducing the ease at which cryoprotectants can be infused into every cell. To solve this problem, an infusion machine was used before supercooling, pumping glycerol and trehalose – two protective molecules produced by supercool insects and desiccation-tolerant animals – at the balmy temperature of a fridge. With these cryoprotectants in place, the organ was then cooled to −4°C in a sealed, liquid-filled bag. Since the air-to-liquid interface is a common place for ice formation (think of the surface of a pond), this reduced the potential nucleation sites.

While transplantation wasn't possible in this early trial, the livers were nonetheless warmed and reperfused with blood. While there was some damage, it was within the realm of any transplant, regardless of how long the organ had been preserved. This study suggests that livers can be preserved by supercooling for 44 hours and still be transplanted successfully, nearly four times the limit using current methods at 4°C.

A supercooled organ is one option. In 2022, the same group of scientists and surgeons from Harvard Medical School tested whether a wood frog's approach to surviving winter might offer even greater periods of preservation. Rather than trying to avoid ice formation completely, Korkut Oygun and his colleagues employed ice nucleators to limit ice formation to spaces in between cells. In order to protect the delicate membranes and internal chemistry, they also infused the livers with a variant of glucose, 3-O-methylglucose (3-OMG). Lowering the temperatures to −15°C, a similar temperature to that experienced by a wood frog in northern Alaska, they found that livers were preserved and able to be transplanted after *five days* of freezing. As with their supercool work, these were rat livers and not the ginormous organs we carry around next to our stomachs. But if replicated in human organs, it might not be a question of whether freeze-tolerant animals can lead to revolutions in medicine, but which ones. Is the future of cryopreservation frog-like or squirrel-like?

While temperatures in northern Alaska swing from temperate to deep freeze, the waters surrounding Antarctica remain stable, hovering between −1.9°C and +1°C. But this is changing. For an ecosystem that has cooled over millions of years, Antarctica is now warming at an alarming rate. For animals that have adapted to extremes of stability, taking comfort in consistency, even small changes can be catastrophic.

Since 1950, the air that circulates around Antarctica has warmed by 3°C, a rate five times faster than the global average. Along the coastline, however, there has been evidence of cooling down to 200 metres, the depth where the majority of animal life – including icefish – can be found. Further out, into the surface waters of the Southern Ocean, however, and there has been a general trend of warming: roughly 0.3°C per decade (3°C over a century). If these trends become the norm, both at depth and near to the Antarctic coastline, it could have huge consequences for animals adapted to stable, subzero waters. Warmer waters contain less oxygen, and the bloodless icefish – lacking oxygen-transporting haemoglobin – may represent an 'evolutionary dead end', says Melody Clark at the British Antarctic Survey.

And this is just the most obvious impact of warmer waters on Antarctic fauna. The stress of other animals is more subtle. Since 2017, Clark and Lloyd Peck, along with Gail Ashton at the Smithsonian Environmental Research Center, have placed dozens of heated plates into the freezing waters of Ryder Bay in the Antarctic Peninsula. When warmed to 1°C above that of the surrounding water, the primary colonizers of these seabeds, filter-feeding bryozoans and spirorbid worms, grew faster and larger. The worms, secreting their snail-like calcium carbonate shells, were 70 per cent larger. From the outside, they were thriving.

But when Clark's PhD student at the time, Leyre Villota Nieva, analysed the worms' internal chemistry, she found that they were flooded with protective chaperone proteins and DNA repair enzymes. They were stressed.

Then winter descended. While the plankton-infused waters of the Antarctic summer permitted higher growth rates, two months of darkness at Ryder Bay led to the annual dearth of plankton – microbes dependent on sunlight.[9] Warmed by just 1 °C, the bodies of the bryozoans and worms demanded more nutrients. A hotter body needs more food than a cold one. Unable to meet demand with supply, they were slowly dying. 'They just haven't built up enough reserves to survive the winter,' says Clark.

Bryozoans and worms are just a couple of animal groups that depend on the amount of plankton in the Southern Ocean. Krill, those shrimp-like crustaceans that swarm into the billions, are the fodder of penguins, baleen whales and the misleadingly named crab-eater seals (a species that filters crustaceans through its teeth and doesn't eat crabs of any sort). The sharp-toothed icefish feed on Antarctic whiting. Sea spiders suck out the innards of their favoured anemones, soft corals and worms. Starfish scavenge along the bottom, cleaning up the dead and decaying. How these intricate food webs restructure in the strong swirl of the circumpolar current is anyone's guess. But it is those creatures dependent on sea ice for their breeding – krill and penguins, in particular – that are most at risk.

In 2023, sea ice around Antarctica was at its lowest level since records began: 15 per cent less ice formed in the month of July (the middle of winter) than the four-decade average. Emperor penguin chicks, still without their waterproof and streamlined plumage, lost their platforms and fell into the frigid waters and drowned. According to a recent threat assessment published in the journal *PLoS Biology* in 2022, emperor penguins are likely to go extinct by the end of this century. That's assuming that these animals would continue to lay their eggs on ice that melts before their chicks have fledged. But emperor penguins, as clumsy as they are, are still adaptable, a modern representative of an ancient family of flightless birds that have swum through our oceans for 60 million years. And recent satellite imagery shows that they are already moving to places of permanent ice in response to a warmer, less icy, world.[10] 'Emperor penguins have taken it upon themselves to try to find more stable sea ice,' Peter Fretwell, a researcher at the British Antarctic Survey, said at the time. 'When we do get future ice losses, emperors can and will move. It's in their nature… this is a species that has to be dynamic.'

Antarctica isn't a static landmass covered in everlasting ice. The images of ice melt and glaciers calving skyscraper-height bergs largely come from the west side of the Antarctic Peninsula, a place where change has been a feature of its history. 'The climate change we're seeing in the West Antarctic Peninsula is perfectly normal,' says Andrew Clarke, a former researcher at the British Antarctic Survey who began his work on cold adaptation in the 1960s. 'It's happened several times in the past. It's driven by oceanographic processes in the Pacific. So it's really difficult [to talk about] because the

West Antarctic Peninsula is one of the classic examples of climate change – you see it everywhere. Everybody quotes it. And yet, much of what we see is probably a natural process. However,' he adds, 'there is some evidence that we've screwed it up a bit. And what we're seeing now has an element of human activity on top of that.'

What is irrefutable is the situation in eastern Antarctica, the place of vast ice plains and the coldest temperatures on Earth. It was once thought to be impervious to climate change, and there were even indications that it was actually gaining ice in recent decades, but now eastern Antarctica has begun to break apart. 'There, you've had really dramatic ice shelf collapses: Larson A, Larson B and Larson C,' Clarke says. 'And those are unprecedented. So it's clear that on the east side of the Antarctic, human-induced climate change is having a dramatic effect.'

With emperor penguins already moving to more stable ice floes to raise their chicks, these icons of the Antarctic are in a race against time. Just as polar bears are moving northwards as multi-year ice melts into annual floes, the plight of the emperor penguin is an hourglass that should be used to spur economies toward net zero carbon emissions.

To say something has frozen often means that it has turned to ice under subzero temperatures, but it is also a synonym for stability – to freeze a moment in time is to keep it forever. But the planet we live on is changing from the tectonic plates under our feet to the atmosphere above our heads.

As I mentioned at the beginning of this chapter, Antarctica hasn't always been an ice world. In the coming millennia, this great southern continent might be returning to a more temperate climate, a place of rock and greenery in places where ice once covered mountains. Emperor penguins might eventually go extinct as the rate of warming outpaces their ability to adapt. Even in the worst case scenario of an ice-free Antarctica, what brings me comfort is to think of the resilience of animal groups, not just their individual species. The penguin family tree encompasses 18 species that live all across the southern hemisphere, from New Zealand to South Africa, Namibia to Argentina. Above the equator, there are colonies of penguins on the Galápagos Archipelago. Expert divers and swimmers, some species thrive in subtropical waters as well as Antarctic ice floes, an insurance policy of evolution.

In 2021, gentoo penguins were spotted breeding within the Southern Ocean for the first time. Requiring bare rock and no ice to lay their eggs, this discovery has been called a canary in the coalmine for climate change: the Southern Ocean has already warmed enough to form an ice-free island for these rock-loving birds. But there's another angle to this story – life is adaptable. There is an innate intelligence built into all living things. Survival isn't simply about finding food and fending off predators; it is the ability to move from an inhospitable place to a hospitable one.

Looking at the penguin fossil record, these birds likely evolved along the coasts of Australia around 60 million years ago, a time when this region was hotter and more humid than today. As the Southern Ocean formed some 30 million years later, sending Antarctica into its phase of extreme cooling, only then could these birds depend on ice for part of their life cycle. Put another way, penguins were here long before the glaciers of ice that now define the landmass of Antarctica. With meaningful actions to reduce our carbon emissions, they will continue to dive into the Southern Ocean for many millions of years to come. It will be a different ocean to that which first cooled some 14 million years ago, but a group of black and white birds will still call it home.

CHAPTER FIVE

HIGHS AND LOWS

Pressure

Pale in plumage and the weight of a well-fed rabbit,[1] a bar-headed goose flies from coastal lagoons in southern India to freshwater lakes in China and Mongolia over 11 weeks. It's a journey of more than 3,000 kilometres, well short of the bar-tailed godwit's trans-Pacific flight of over 11,000 kilometres. It has nonetheless been called the most spectacular athletic performance of any animal, for one reason: the Himalayas lie in the middle of this journey, the 'most formidable mountain range in the world'. The average height of these peaks is around 4,500 metres,[2] and there are 14 mountains that reach above 8,000 metres, the so-called death zone of mountaineers. But these inhospitable peaks don't seem to be a particularly difficult challenge to a bar-headed goose, as mountaineers scaling Mount Makalu, the fifth tallest peak in the world, witnessed first hand in 1954. 'At 16,000 feet, where I breathed heavily with every exertion and where talking while walking is seldom successful, I had witnessed birds flying more than two miles above me, where the oxygen tension is incapable of sustaining human life – and they were calling,' the high-altitude biologist Lawrence Swan wrote in *Natural History Magazine* in 1970. 'It was as if they were ignoring the normal rules of physiology and defying the impossibility of respiration at that height by wasting their breath with honking conversation.'

It's a migration so extraordinary that it's worth repeating: these birds take flight at sea level and fly over the tallest mountain range in the world, all in eight hours. As the editors of the 2017 textbook *Bird Migration Across the Himalayas* wrote, 'We... have been struck with awe every time we were observing these birds doing the nearly impossible.'

The impossibility of flying at extreme elevation is due to hypoxia, a deficiency in oxygen. But unlike the turtles and naked mole-rats of Chapter 2, the proportion of oxygen that surrounds a bar-headed goose never changes, whether it is taking off from an Indian mudflat or flying past Mount Everest. The atmosphere at high altitude is simply more spread out, diffuse. Distant from the pull of Earth's core, the molecules of oxygen, nitrogen and carbon dioxide are further apart. In other words, due to the low hypobaric pressure at altitude, air is thinner, and a lung can't suddenly increase in size to accommodate the change.

The low pressure also has an impact on the mechanics of flight. With fewer molecules to push against, lift is harder to achieve and maintain. 'At 8,000 metres,' one study into bar-headed geese published in 2012 notes, 'the minimum mechanical power required for flight is 50 per cent greater than that at sea level.' To add a little more oomph to each wingbeat of their migration, these birds fly at night or in the early morning, a time when air temperatures can be as low as −30°C and the cold brings molecules slightly closer together, just a smidge. Just as cold butter is firmer than warm butter, cold air is thicker than warm air. Freezing cold, with only the stars above them and the wind often blowing down the mountains into their faces, bar-headed geese seem to fly straight into adversity.

As anyone who has scaled a mountain can attest, whether it is Ben Nevis (1,345 metres) or Kilimanjaro (5,791 metres), the primary stress that comes with altitude is feeling out of breath. If we remain at altitude for a long time, or live there permanently, our bodies also start to produce more haemoglobin, the oxygen-transporting molecules in our blood. While this helps grab hold of the fewer oxygen molecules in the air, it also comes at a cost. 'If you have high levels of haemoglobin, your blood tends to be more viscous, and that can have a lot of damaging effects,' says Tatum Simonson, a geneticist who studies adaptations to high altitude. 'You're basically pumping this very thick, concentrated blood throughout your system. Your heart is on overdrive.'

One potential outcome of this added stress on our circulatory system is chronic mountain sickness (CMS). First described in 1925 by the Peruvian doctor Carlos Monge Medrano, CMS (also known as Monge's disease) can afflict people who have lived happily at high altitude for years. 'It's not clear what triggers the onset,' says Cynthia Beall, an anthropologist who has studied high-altitude populations in the Andes and Himalayas since the 1970s. 'But people become breathless, they become cyanotic [lips and extremities turning blue], they can't work, they can't sleep well – they're very ill.' According to a paper from 2014, 'The most frequent symptoms and signs of CMS are headache, dizziness, breathlessness, palpitations, sleep disturbance, mental fatigue, and confusion. People affected by CMS often suffer from stroke and myocardial infarction [i.e., a heart attack] in early adulthood, mostly due to increased blood viscosity and tissue hypoxia.' As with short-term altitude sickness, the remedy for CMS is a slow descent into thicker, more oxygenated air. But this is no cure. Fluid may have already built up in the lungs (a high-altitude pulmonary edema, or HAPE) or in the brain (a high-altitude cerebral edema, or HACE). Even in villages and towns situated high in the Andes, 10 to 20 per cent of men (who are at higher risk of altitude sickness) develop CMS at some point in their lives.[3]

By acclimatizing our bodies to gradual increments in altitude, however, the human body can be trained to function at extreme altitudes. Our bodies produce more red blood cells and more haemoglobin to grab hold of whatever oxygen is available in the atmosphere. With the correct preparation and training, even Mount Everest – once thought to be inhospitable to our bodies, the air at 8,849 metres too thin to power our brain and muscles – can be surmounted without supplemental oxygen. In 1978, the Italian mountaineer Reinhold Messner was the first to reach its summit and inhale. 'I am nothing more than a single narrow gasping lung, floating over the mists and summits,' he wrote. Over the next 16 years, again without supplementary oxygen, he would reach the top of all 14 peaks over 8,000 metres.[4]

As mammals, we breathe tidally, inhaling and exhaling separately just as the ocean rolls up and down the shore in a regular and repeated pattern. This makes the thin air at high altitude a particularly difficult obstacle to overcome. We can either breathe more or increase tidal volume, but there is always time when the air in our lungs isn't being replenished (just before we exhale). Birds, however, don't struggle with the same breathlessness as mammals. No matter the altitude, they don't seem to suffer from HAPE. While we breathe in and out, our lungs expanding and contracting, birds breathe in one continuous circuit. Their lungs are rigid, only moving a fraction during each breath. But they are filled and fuelled by eight or more balloon-like air sacs that inflate with every breath and, in sequence, squeeze air over the lungs' surface. While our lungs are dual-purpose – sites for gas exchange and mechanical breathing – birds have separated these two functions. They breathe with air sacs that don't provide a surface for gas exchange. And their lungs are largely inflexible, organs that only have to worry about getting oxygen in and carbon dioxide out.

Put together, birds can maintain more air flow over their lungs' surface than mammals. Instead of breathing quickly to replenish air in their lungs, birds breathe longer and more deeply. Their trachea is four and a half times larger than a similarly sized mammal. 'These specialisations appear to have permitted birds to inhabit some of the most extreme and varied environments on earth,' Sabine Laguë from the University of British Columbia wrote in 2017. Hoopoe larks live in the scorching heat of the Arabian Desert. An ostrich can sprint through the African savanna at 70 kilometres per hour on two legs. A house sparrow can breathe easily at simulated altitudes of 6,100 metres; a house mouse would be comatose under these same conditions. While our (mammalian) breathing is as languid as the ocean's tides, the avian system is more like a racetrack.

Bar-headed geese have refined this racetrack into one of life's most efficient courses. The haemoglobin inside their red blood cells can grab hold of oxygen molecules more efficiently than most other birds, a feat that comes from a few unique mutations in their DNA. Their blood vessels are incredibly thin, two and a half times smaller than our own, making the journey for oxygen and carbon dioxide much smaller and, therefore, faster. Their muscles are surrounded by twice as much capillary as a mammal's. Inside their cells, the mitochondria, those tiny power factories of complex life, are positioned right next to the surface where oxygen comes in, eagerly awaiting its arrival. These minute, often nanoscopic, changes have allowed these birds to endure the most inhospitable heights without any prior training or acclimatization.

The highest flying bird is not a bar-headed goose, however. That record was set in 1973 by a Rüppell's vulture that collided with a commercial aeroplane 11,300 metres above the city of Abidjan, Ivory Coast. Like eagles and buzzards, these scavengers soar on the hot air rising from sun-scorched rocks or bare soil below. Like bar-headed geese, these vultures have mutations in their haemoglobin genes that have allowed them to breathe easily at an aeroplane's altitude, when they're not getting pulverized by jet engines, that is. While the plane landed safely using just one engine, the bird was only identified using a few wing feathers.

But bar-headed geese don't soar. Since 2008, Lucy Hawkes,[5] a physiologist at the University of Exeter, and her colleagues have strapped data loggers onto these birds using specially designed harnesses. Monitoring their heart rate, temperature and altitude, these Fitbit-sized pieces of tech have shown that these birds keep flapping, no matter what the environment is doing. Tailwind or headwind, they power through. With no thermals to float on, they follow the rise and fall of the land below. Even after they have flown over the Himalayas and entered the flat Tibetan Plateau, the so-called roof of the world, they don't take it easy. Flying in V-formation and honking clouds of steam into the frigid air, bar-headed geese don't glide. 'They keep flapping,' Hawkes says. 'They love it.'

While observations from mountaineers suggest that these birds fly over Mounts Everest and Makalu, Hawkes's data loggers have never recorded such a high-flying goose. From a dataset of 91 geese, the highest altitude was 7,290 metres, more than a kilometre off the summit of the highest peaks. The majority of geese flew between 6,000 and 7,000 metres. Even so, in previous work with captive birds in wind tunnels, researchers in the late 1970s found that bar-headed geese seem unfazed by a simulated altitude of 12,000 metres. While they weren't flying in these experiments, they were still actively walking, suggesting that their limit is well above the 7,290 metres that Hawkes and her colleagues have recorded.

Even taking this conservative altitude as their maximum, why these birds fly to such elevations is still a mystery. As two physiologists wrote in 1980, 'there must be a good explanation for why the birds fly to the extreme altitudes... particularly since there are passes through the Himalaya at lower altitudes, and which are used by other migrating bird species.' Do they avoid predation by steppe eagles by flying away from canyons that could block their escape? Does a higher elevation provide a better guide for their migration, the Himalayas unfolding beneath them like a map? In his 1970 essay in *Natural History* magazine, Lawrence Swan added that flying at 9,000 metres might be advantageous when the peaks are swept by 200 mph winds of the jet stream, a blast of hurricane-force winds that could push even the most powerful of flyers into the mountainside. While 'the summit pyramid was boiling with white spumes of clouds' and the ridges and canyons below were filled with 'twisting vortices of snow, like miniature tornadoes', a bird could fly in the calmer air above this maelstrom. It would still be buffeted by stratospheric winds, 'driven hundreds of miles out of its course', Swan adds, 'but it could survive'.

An alternative theory, one that Hawkes and her colleagues wrote about in 2015, is that these birds have been migrating over similar distances long before the Himalayas had grown to such world-beating heights. '[T]he species (or its ancestor) may have begun migrating between South and Central Asia in the late Pliocene or Early Pleistocene… a time in geological history when the Himalayas were not nearly as high,' they wrote. Birthed by the collision between the Indian subcontinent and the landmass of Asia some 50 million years ago, these windswept peaks have been forced ever higher by the gradual crushing of tectonic plates below. These birds are flying to extreme heights because their ancient flightpath has slowly been pushed higher. 'Surely a mountain range reaching that high through the atmosphere into the frigid lower limits of the Stratosphere would defeat the flight of birds?' Swan wrote in 1970. 'No – the birds beat the mountains.'

The study of mountains helped found one of the central principles of ecology, the study of how animals behave and where they live. Beginning with the work of Alexander von Humboldt in the late eighteenth and early nineteenth century, it became apparent that animal life wasn't randomly rolled out over the surface of the Earth – there were general patterns. These could follow the lines of latitude – more biodiversity in the tropics than the poles – but also altitude.

The higher you walked above sea level, as a general rule, the fewer species there were. This found its most extreme confirmation on the tallest peaks. Writing in 1840, a few decades after the foundational work of Humboldt, the British zoologist Edward Forbes noticed the marked decline in his favourite study animals – snails and other molluscs – with altitude. As he walked up the tallest peaks of England and Wales, biodiversity soon dwindled to one remaining species: *Oxychilus alliarius* or, more commonly, the garlic snail, so named for its scent when disturbed. Only this stinky little snail could tolerate 'the exposed conditions near mountain summits'.

From his early work into mountain ecology, Forbes then turned his mind to the opposite of altitude: depth. This allowed him to study his most beloved of molluscs: aquatic snails. Dredging the seabed surrounding the British Isles and as far away as the Mediterranean, he thought that their diversity would decrease with depth, just as it had with altitude and land snails. 'The parallel between terrestrial and marine systems was obvious,' two scientists from the National Oceanography Centre in Southampton wrote in 2006, 'extreme environments, be they on the tops of mountains or at the bottom of the sea, would surely lead to extinction of life forms.'

The first step of science is hypothesis, conjuring up a prediction and then testing it with experiment or observation. And testing his prediction in the Aegean Sea, Forbes found his samples to support his work into mountains. Trawl nets that were dropped to below 350 metres came back full of mud and little else. Published in 1860, his work was read by a scientific community already primed for a denuded deep sea, a place where pressure, cold and lack of food fashion a vast inhospitable zone for complex life. 'The surface of the deep ocean teems with animal life,' one biologist wrote in 1834, 'and this is probably not [the case] down to depths where, from the want of necessary conditions, it ceases to exist.' At this time, humans could only dive as far – and for as long – as a glass-bottomed bucket of air would allow. There were no submersibles, no scuba tanks, no electric lights that could penetrate the darkness below 200 metres. Even the light-filled surface of the ocean was barely accessible to us; how could the deepest parts be hospitable for animal life?

Forbes's research in the Aegean, published in 1844, supported the so-called Azoic Hypothesis (azoic means 'without life') for the deep sea. With this study to turn to for support, the great biologist Louis Agassiz was confident when he wrote in 1851, 'The depths of the ocean are quite as impassable for marine species as high mountains are for terrestrial animals… Not only are no materials found there for sustenance, but it is doubtful if animals could sustain the pressure of so great a column of water.' While the lack of food and fridge-temperature cold were seen as contributors to an animal's demise, it was the unimaginable pressures of the deep sea that became the critical limiting factor. 'According to experiment,' the geologist David Page wrote in 1856, 'water at the depth of 1000 feet is compressed 1/340th of its own bulk; and at this rate of compression we know that at great depths animal and vegetable life as known to us cannot possibly exist – the extreme depressions of seas being thus, like the extreme elevations of the land, barren and lifeless solitudes.'

This hypothesis was so pervasive, so ingrained in the academic zeitgeist, that even ample evidence to the contrary wasn't strong enough to turn the Azoic (the apparent lifeless depths) into the Zoic (the biodiverse abyss). Lines dropped 700 metres down in the Southern Ocean brought up living coral and marine invertebrates. While sounding the seabed for trans-Atlantic cables, a line was brought up from 2,305 metres, a depth well into the midnight zone where light never touches, and had 13 starfish attached to it. In the Mediterranean, the very sea that Forbes was sampling, a telecommunications cable was retrieved from 2,195 metres between Sardinia and Bona and was covered in corals, oysters and clams. As one zoologist aboard the HMS *Bulldog* wrote in 1862, this research 'points to the existence of a new series of creatures peopling the deeper abysses of the ocean'.

Only in 1870, when the Royal Society's own scientists started to sample the deeper realms of our oceans, did the Azoic Hypothesis finally fall. It turned out that, by sampling the Aegean, Forbes was unintentionally using an outlier to build a general rule; this sea is nutrient poor and, therefore, lacking the diversity of life found in the oceans. And then there was the problem with his sampling tools. Loosely woven nets dragged along the seabed using a couple of metal rods weren't exactly windows into the unknown. They tended to be filled with mud very quickly, leaving no room for any animals that dig, indeed live, at these depths. While a bad worker never blames their tools, a good oceanographer might still get away with it.

Just as mountains have dramatic shifts in flora and fauna from their foothills to the tree line, the Alpine tundra to the snow-capped peaks, the oceans were long separated into four zones – sunlit (down to 200 metres), twilight (1,000 metres), midnight (4,000 metres) and abyss (4,000 metres and below) – that were largely defined by how much sunlight they receive. By the 1950s, however, some of the first deep-sea trawlers brought up animals from ecosystems that, some argued, deserved their own category. 'It became clear back in 1954–1956 that the fauna in the deep ocean troughs was so unique that the depths over 6–7 kilometres should be isolated into a special zone in the system of vertical biological zonality,' the Soviet oceanographer Georgii Beliaev wrote in 1989. While the Soviets called it the 'ultra-abyssal zone' and Western scientists called it the 'hadal zone',[6] the one thing they agreed on was that it was a place of 'great originality'.

To learn more about the animal life that calls the hadal zone home, I travelled to SUNY Geneseo, a university in the northeastern, and very rural, reaches of New York State, where I was greeted by Mackenzie Gerringer. She led me to her laboratory on the first floor of the Biological Department, the place where some of her latest samples are preserved in glass jars. 'I think people have this picture of deep-sea fish in their minds, of being very toothy,' Gerringer says.[7] 'You know, [fish] having teeth so large that they have to have grooves in their skull. But when you get to the bottom of where fishes live, you're left with this little pink, transparent fish that's got see-through skin. When you're holding one you can look through the skull and see its brain.' She shows me one of these fish inside a specimen jar. A little wrinkled from the dehydrating effects of ethanol, it was collected by a suction sampler (imagine a vacuum attached to a robotic arm) at 4,000 metres off the coast of California. 'I know they don't look like it, but this is actually one of the best-preserved specimens,' Gerringer told me in a follow-up email. Holding the glass jar in my hands to take a closer look, it is remarkable just how fragile this animal looks. No armour. No toothy grin. Just a gelatinous, tadpole shape that has made the crushing depths of the ocean its home.

One thing that defines much of the deep-sea fauna is the banality of their common names. With names like cusk eels, eelpouts and rat-tails,[8] the denizens of the deep don't capture the imagination like great white sharks or killer whales. And the ghostly fish inside this glass jar was no different. The deepest known fishes on Earth are called snailfish. With representatives living in shallow waters, the name comes from a modified, cup-shaped fin on their undersides that allows them to cling to rocks in the fiercest torrents of water. Adding to this snail-like stability, they also curl their tails around their bodies to complete the look, a fish masquerading as a mollusc. Out of the 400-plus species of snailfish, only a dozen or so have conquered the hadal zone, most famously the Mariana snailfish (*Pseudoliparis swirei*, discovered and described by Gerringer, Thomas Linley and their colleagues in 2014[9]) and the ethereal snailfish (of *Blue Planet II* fame). The latter has elongated, wing-like fins at the side of its body, a delicacy of form that seems so out of place eight kilometres down in the ocean, like a wedding dress worn on the frontlines of war.

But it is an – at the time of writing – unnamed species that holds the record for the deepest fish on Earth.[10] Discovered in August 2022, this species of *Pseudoliparis* (the same genus to which the Mariana snailfish belongs) was caught on high-definition camera at 8,336 metres in the Izu-Ogasawara Trench, southeast of Japan. Snailfish, or Liparidae, in sum, are a group with many shallow-water species, some of which are happiest in the intertidal zone. They couldn't be further from the deep without being permanently on land. As Alan Jamieson, a leading deep-sea biologist who studies the hadal zone, puts it, 'They're the deepest fish in the world but they're not a deep sea fish.'

As a family, snailfish might seem out of their depth living in the hadal zone, but their evolutionary history may also have predisposed them to thrive in places of intense pressure. The latest genetic studies estimate that hadal snailfish emerged around 20 million years ago, perhaps from an ancestor that lived near the South Pole. As Antarctica began to cool at around the same time (as we learned in the previous chapter), this may have provided a preadaptation for snailfish: cold temperatures and high pressures have similar effects on biological material. Both make many proteins and cell membranes stiffen and clump together. Cell membranes, the biological envelope that is made from fat molecules, become rigid – like butter in a fridge. Organisms that live in the deep sea can maintain some fluidity in their membranes by adding more unsaturated fatty acids, molecules that have a bend in their structure. While saturated fatty acids are straight and can be compressed into a tighter space, the bent, unsaturated fatty acids maintain some space between each other.[11] Originating in cold waters, therefore, may have made the proteins and membranes of snailfish more flexible, better able to cope later with the intense pressure of the hadal zone.

This is just a hypothesis. Gerringer tells me that the fossil record of snailfish is notoriously piecemeal. Hadal snailfish may have evolved in cold waters but we don't know for sure. But one part of their physiology that is undoubtedly beneficial to living in extreme pressures is the lack of a swim bladder.

There's a common misconception that bringing a deep-sea fish up to the surface makes them explode. Thankfully for fishermen who regularly catch fish well below the sunlit zone, the guts and other organs of these animals stay within their skeletons. Except for the swim bladder. This organ that is used for buoyancy, inflating and deflating in order to rise and fall in the water column, is less like a bladder and more like a balloon. For a deep-sea fish brought up a few thousand metres – through several hundred atmospheres – this balloon starts to expand under the reduced pressure. Brought up on deck in a trawl, lobster pot or seine net, the swim bladder erupts from the mouth like some gruesome speech bubble.

Causing a grisly death when brought to the surface, the swim bladder also limits how deep a fish can dive. As any submersible design shows, the deep sea abhors a cavity of air. That's why deep-sea submersibles are spherical and very thick, holding back the weight of water from all sides. Pockets of air plus intense pressure have a predictable outcome: implosion. A simple example would be a jam jar dropped into the Mariana Trench. Without a lid, the glass would drop right to the bottom of Challenger Deep, the deepest known point of the Earth's seabed, without changing in shape or form. Screw the lid on, however, and it would implode long before it reached the hadal zone, the force of water crushing the metal lid and shattering the glass. While there's no conscious intent in a body of water, it seems to be in constant battle with air, a foreign state of matter sent from the skies above.

Snailfish don't have swim bladders. When they are brought up from the deep sea, Gerringer tells me, they don't explode or have organs hanging out of their orifices. They look a little worse for wear, largely because they are gelatinous and usually supported by intense pressures. When patrolling the deep for their crustacean prey, however, they have the place mostly to themselves, the cusk eels and eelpouts – and their swim bladders – only occasional visitors to the hadal zone below 7,000 metres.

Snailfish are in a league of their own when it comes to deep-sea fish. But even Liparidae have their limit. The *Pseudoliparis* recorded at 8,336 metres, for instance, may actually be very close to the deepest fish we will *ever* find. 'I don't think there's any way in which we'd find a fish 1,000 metres deeper or even maybe *100 metres* deeper [than this],' Jamieson, the lead scientist on the research project said in 2023. 'We're confident now that we've really really understood this.'

The 'this' he was referring to was the depth at which fish can live, the place where oceans become inhospitable even to the most adventurous snailfish.

In a paper published in 2014, Paul Yancey, along with Gerringer, Jamieson and a couple of colleagues, proposed that no fish can live permanently below 8,500 metres. The reason isn't for a lack of food – there are vast numbers of amphipods below this invisible line, largely because they are free of predation. This lower limit for *vertebrate* life is set by the way they have evolved to cope with the crushing pressures at these depths: inside their cells, a molecule known as TMAO prevents the cell from being crushed.[12] Just as antifreeze proteins prevent water molecules from forming an ice lattice, TMAO holds water molecules away from proteins and membranes inside a cell, a type of biochemical referee that keeps boxers to a safe distance. With some wiggle room to bend and flex, deep-sea animals are as dependent on TMAO as we are on our rigid skeletons.

Since he began this study into TMAO in the 1990s, Yancey has found that the deeper the fish lives in the ocean the more TMAO it packs into its cells. The straight line on his graph – depth with TMAO concentration – goes down through cusk eels to rat-tails and, more recently, to snailfish. But at 8,500 metres, he noticed, the amount of TMAO needed to counteract pressure would hit a new threshold: packed full of 'osmolytes', the fish's cells would become more concentrated than the surrounding seawater.[13] 'The osmotic gradient would reverse,' Yancey says. 'That would cause water to come *in*. And marine fish kidneys can't deal with that; they have no mechanism for getting rid of water, they can only get rid of salt.' A fish at these depths would bloat, its proteins would malfunction, and if it didn't swim upwards, it would die.

Yancey adds that his hypothesis is there to be tested. In lectures to his students at Whitman College in Walla Walla, Washington State, he uses the precedent of Edward Forbes and the Azoic Hypothesis of the deep sea. 'He was a good scientist, as far as I could tell. He wasn't trying to say *this* is the way it is, but people took it like that.' Who knows, TMAO might not be the limit that Yancey has proposed. Fish may go much deeper. With an increasing number of cheaper seabed landers, ROVs and billionaires' submersibles, it is being tested to ever greater acuity with each passing year. 'If you find a stray fish at 9,000 metres, [that] isn't going to disprove anything,' Yancey says. 'You'd have to find a [population] that lives there. Then I would be clearly wrong.'

Before this inhospitable line at 8,500 metres or so, snailfish and the waters of the deep ocean are harmonious – a slow-moving, gelatinous body and the deep ocean in perfect osmotic balance. Pink, translucent, the size of a pond-dwelling goldfish, a hadal snailfish swims through 800 atmospheres without any threat from predation or competition. It sounds like a sweet deal, an evolutionary gamble that paid off.

This would be true if it weren't for the earthquakes.

After showing me her specimens of snailfish and a CT-scan of the strong secondary jaws that lie in the throats of these otherwise flimsy fish, I begin to nose around Gerringer's lab at SUNY Geneseo. A strange map of the world on the wall catches my eye. As with any atlas, this one shows the continents and the oceans. But there are no countries on land, no bright colours or borders. The Americas, Europe, Asia, Africa are all shaded in grey. This is a map of the oceans' borders, the fault lines where tectonic plates meet.

Along black lines, red dots show magnetic anomalies and earthquakes. Where they accumulate into a mass of red, these are the subduction zones – the trenches. The edges of the Pacific Ocean, whether in the east or the west, are covered in a rash of dots. 'Trench, trench, trench, trench,' Gerringer says as her finger traces the tectonic plate's edge from the west coast of South America up to Alaska, Russia, and back down the eastern coasts of Japan and into Oceania. While trenches are found all over the world, this is where the world's most active and deepest trenches are found. All nine of the 10,000-metre-plus trenches are found where the Pacific plate is pushed underneath the continental plates of Russia, Japan and Australia.

Another misconception of the deep sea is that it is stable and boring. While this might be true for some parts of the vast abyssal plains of mud, the hadal zone is a riot of activity. One study of the Mariana Trench, for example, recorded 2,000 earthquakes in a 90-day period, Gerringer tells me. 'Not all of those are huge,' she adds. 'But these are much more active sites than we used to think.' While there are fractures where molten rock can burst to the surface and form 'rock pillows', trenches are mostly characterized by enormous mudslides. One underwater mudslide off the coast of Japan in 1972 was estimated at 720 kilometres (in volume, not area), all falling in one cataclysmic event. For comparison, the eruption of Mount St Helens in 1983 created a mudslide of 2.5 kilometres (288 times smaller).

This geological activity is recorded in the bones of hadal snailfish. While on deck in the open sea, Gerringer used a sharp scalpel to extract the ear bones of dozens of specimens that had been caught in a trap that had been dropped

7,000 metres into the Mariana Trench. Gerringer wanted to know how old it was. While other deep-sea fishes, such as the commercially harvested orange roughy, live for a century and don't reach maturity until they are at least 20 years old, the life cycle of a hadal species was little known. Some of the first nets that scoured the deepest trenches in the 1950s had caught snailfish but, since they were trawled, they were not in good condition. The modern box traps that were dropped and closed without dragging along the seabed helped to preserve their condition. After she had found the largest, chia seed-sized ear bones, or otoliths,[14] Gerringer then polished them down to their centre and placed them under her microscope. Just as a dendrochronologist can estimate the age of a tree from its rings, the number of lines in these otoliths can be used as a guide to a fish's age when it was caught. Even in the deep sea, the flow of nutrients coming from the sun-dappled waters above changes with the seasons. This change in food abundance affects a fish's growth rate, etching a physical mark into its ear bone. The change from a harsh winter to a bountiful spring is written in bone, a palimpsest of productivity. After polishing 66 snailfish ear bones (a phrase you don't hear enough), she found none of the estimated ages were above 20, a fifth of the age of an orange roughy.

Snailfish, as far as deep-sea life goes, live fast and die young. They grow, they mature, they reproduce, all before a cascade of mud buries them and they get recycled by hungry scavengers. They live in a realm largely oblivious to our devastating impacts on the natural world. Their main threat is a shift in tectonic plates, and the biblical-scale mudslide that it generates.

Mountains have carved bar-headed geese into high-altitude athletes. The snailfish has been shaped by the constant return of the Earth's crust into magma.

There is animal life below 8,500 metres, below the deepest snailfish. Sea stars, isopods, sea cucumbers, glass sponges: all have representatives that filter water or sediment to feed in waters over ten kilometres down. (Unlike fish, marine invertebrates can match their cellular osmolarity to the water they live in. They are known as osmo-conformers, and adding TMAO to protect themselves against high pressure can simply be switched for any other osmolyte, such as free amino acids (FAAs) and other methylamines.) But if there's a dominant feature of trench life it is amphipods, the scavengers that snailfish most often prey upon. Where there are no fish predators, however, amphipods swarm. A place that is inhospitable for fish is very hospitable for amphipods. The only thing that can eat an amphipod below 8,500 metres is another, larger amphipod.

Send a piece of mackerel down into the deepest trenches and it will soon – within minutes – be covered in amphipods. Most of these are no larger than your fingernail, but some – the supergiants – grow to the size of sewer rats. Together, they can strip a fish to the bone in a couple of hours. After these scavengers have found the bait, the predators soon arrive: elongated amphipods with long, oar-like legs that feed on the smaller scavengers. They arrive so quickly that deep-sea scientists are often stumped as to where they came from. Were they buried in the mud nearby? Do they travel long distances along the scent lines of dead fish? To a static camera illuminating a few metres of the hadal zone, it looks like they have the ability to teleport.

While an image or video full of anaemic crustaceans isn't going to spark headlines around the world, the ease at which amphipods are collected is a huge boon for deep-sea biologists. 'We've learned a ton about the hadal zone through them,' says Johanna Weston, a deep-sea ecologist who started working with hadal amphipods as part of a PhD at the University of Newcastle. By analysing and comparing their DNA, she can learn more about how connected (or isolated) distant trenches are, a possibility that indicates that the most remote trenches in the world are the Puerto Rico or Peru–Chile; the amphipods there are more dissimilar than any others. By looking at amphipod gut contents, she can understand how the deep sea is connected to the water above. And by learning more about their diversity, a picture emerges that life in the ocean's trenches doesn't just survive. It thrives. 'We just don't have the numbers yet for other [animal] groups to make such huge advances,' Weston says.

When I visit Weston at WHOI, she brings a couple of glass vials in her backpack. They each contain an amphipod previously unknown to science. Collected from the bottom of the Puerto Rico Trench in 2016, she lifts them out of their clear preservative and places them on a Petri dish on the picnic table we're sitting at. The smell of strong ethanol tickles my nostrils even though we are sitting outside. I look down at these animals from a distant land and think of how best to describe them. One amphipod is greyer, rounder and has shorter legs than the other. And then I run out of ideas. They don't have claws like crabs or lobsters. Their antennae are tucked into their mass of legs, at least in this preserved posture. They look like skinny, anaemic woodlice wearing stilts.

Weston sees them differently. 'I, personally, think they're beautiful,' she says. Depending on what they've eaten, their shells can shine in autumnal colours of rust orange to sunset pink. 'There are other ones that just have crazy long legs or a really elongated body so that they're really strong swimmers and really good predators.' Still, Weston acknowledges that her adoration didn't come naturally. 'Before I started my PhD, I didn't know what an amphipod was.' Faced with a cupboard full of these crustaceans from around the world, she knew that there was more to them than meets the eye. While well-preserved fishes were scarce or absent altogether, these animals were portals into the deepest trenches. 'You've just got to decide: I'm going to think these are the coolest things ever,' Weston says. 'And I want to think that they're super cute because I've got to look down a microscope [at them], a lot.'

Through her PhD research, she even helped make amphipods global icons. While she was dissecting a juvenile amphipod collected from 7,000 metres down in the Mariana Trench, Weston found a blue microscopic fibre inside its stomach. Just over half a millimetre long and shaped like an archer's bow, it was a sliver of polyethylene terephthalate: PET, or polyester, a type of plastic used in clothing and food packaging. In particular, it's the type of plastic that is used to make water bottles. In an attempt to shine a light on the far-reaching impact of our consumer choices on land, Weston named the species *Eurythenes plasticus*. Along with her supervisor, Alan Jamieson, she wrote, 'This name speaks to the ubiquity of plastic pollution present in our oceans.'

There's a misconception that the deep sea is so far away that it is pristine. While the depth of the Mariana Trench is often described as Mount Everest plus another few thousand metres, this comparison forgets the important difference of gravity. To get to the top of Everest, something has to be carried or blown up. To find itself in the Challenger Deep, a plastic bottle simply has to sink. One 2020 study into microplastics found in the hadal zone called deep-sea trenches the 'ocean's ultimate trashcan'.

There are an estimated 250,000 tonnes of plastic floating on the surface of our oceans. But it is only one type of trash. Combining the US government's National Oceanic Atmospheric Administration footage with her dive in the submersible *Alvin* in 2022, Gerringer has found hundreds of obscure and troubling items in the ocean's depths. 'Things like an old torpedo, but also jeans, hot sauce packets, cans, scrunchies, a ladder, all sorts of stuff,' she says as we stand in her lab, a poster of a few of these examples pinned up in the corridor outside. 'It's ending up in these habitats, so they're not untouched. They're not otherworldly, they're very much integrated with the rest of the ocean.'

After Yancey has told me about the potential limit for fishdom, the 8,500 metres line where seawater would be absorbed by fish, I ask him whether he thinks that fish might ever break through this barrier. I recalled a quote from the palaeobiologist and author Thomas Halliday, someone who looks back in time to study evolution today. 'The fossil record shows us time and time again that, whenever a new niche opens up, whenever there is a new resource to exploit, something evolves to exploit it,' Halliday writes in his 2022 book *Otherlands*. 'Nature is nothing if not inventive.' As the current era is forever passing deeper into history, I wondered whether millions of years of evolution would find a way to break through this osmotic limit.

There are definitely a lot of amphipods below 8,500 metres, Yancey tells me. Not only does this add some indirect support for his theory in the sense that snailfish aren't eating them at this depth and so they can swarm with impunity, it also offers an untapped resource. The only thing missing from the basic equation of what drives evolution into new niches is competition, or a lack of it. Fish at 8,000 metres, whether it's a snailfish or a wandering cusk eel, have the place pretty much to themselves. If this situation changes in the distant future, if eelpouts or rat-tails lose their swim bladders and descend further into the trenches, adding more competition into this niche, then perhaps there will be an added push to live deeper.

It would require drastic changes in their physiology, however. A snailfish's gills and kidneys would have to pump water out of their blood, not salt. It would be the opposite of what a marine fish has evolved to do. But it isn't impossible. Salmon do this as they switch from their oceanic lifestyle and start to migrate to their breeding sites in rivers and streams far inland. It takes a few days for their bodies to adjust, but they transform from marine fishes into freshwater ones. 'Fish can do it,' Yancey says. 'But only those that have evolved to it over time.'

In another 50 million years, who knows, there could be a snailfish in Challenger Deep, soaking up water from its surroundings and flushing it out of its system just as we excrete urea in urine. If it rises above 8,400 metres, it might change its gills and kidneys back to a salt-extracting process. But why would it need to? They have all the amphipods they could ever eat down there.

CHAPTER SIX

LIFE IN THE FURNACE

Heat

The ant drops its cargo and cleans its antennae, rubbing each one through its mandibles in turn. I'm trying to focus the lens of my camera as it does so, my bare knees pressing into the coarse and very dry earth. Anyone watching must wonder what I'm doing here, kneeling among a patch of dried thistle, cow parsley and the occasional discarded beer bottle, trying to capture a portrait of an animal that few people would stop to look at, never mind photograph. As the entomologist E.O. Wilson once wrote, 'ants are everywhere, but only occasionally noticed'. But I haven't just stumbled upon this area, I was brought here by Xim Cerdá, a researcher at the Biological Research Station in Seville, southern Spain, someone who has studied these ants for more than 20 years. This particular area – a patch of dirt just off the motorway that leads south from the city – is one of his study sites. The ant that I'm admiring through my lens is *Cataglyphis velox*, just one member of a group of ants that are most often associated with scorching sand dunes in the Sahara Desert and the baking salt pans of Tunisia. *Cataglyphis* ants are heat-tolerant, i.e., thermophiles (or 'heat-loving').

Along with two distantly related families of ants that live in Australia and Southern Africa, these insects sit at the upper limit of what's possible for an animal living on land. The individual I'm admiring can continue to forage when the ground is over 60°C. In fact, they don't emerge from their underground nests until the heat has forced other animals into their shady refuge. This can be around 40°C or higher, a level of heat that causes the proteins they are built from to unravel into inaction. In this inhospitable heat, *Cataglyphis* ants erupt into a world other animals hide from. As Rüdiger Wehner, the doyen of *Cataglyphis* ants, once wrote, 'all foragers leave their underground nest in an explosive outburst confined to a few minutes per day during the hottest midday period'.

The worker ant I'm watching is carrying another ant in its mandibles, a less thermo-tolerant species whose hours of activity are focused within the cool of morning and evening. *Cataglyphis* ants are known as thermal scavengers, subsisting on those insects that succumb to the midday heat. While better-known scavengers such as hyenas or vultures thrive on the kills of predators, these ants subsist on the sun's daily victims: a fly that didn't find shade fast enough; an ant that got lost and couldn't find the cool burrow that led to its nest; a beetle that, Icarus-like, flew too close to the sun.

The hotter it gets, the more death there is to be found. This general rule means that *Cataglyphis* ants are pulled towards their own demise in order to feed the colony. Although thermo-tolerant, they still have their own limit. If the body temperature rises above 50°C or 55°C, depending on the species, they might become just another dried-up body to be found. In fact, the lengths these ants go to find food once had biologists stumped. 'It is, however, surprising that organisms should subject themselves to near-lethal temperatures in order to maximize their intake of food,' one author wrote in 1985. To put yourself in such a dangerous situation doesn't seem like an evolutionarily successful strategy. But, it turns out, ants may be particularly suited to this risky lifestyle. Eusocial animals with a breeding queen and castes of cleaners, soldiers and foragers, the ants roaming through the midday heat are sterile and will never breed. As with the naked mole-rats we met in Chapter 2, their deaths wouldn't hugely affect the reproductive success of the colony. As long as most of the ants are successful in finding food for their nestmates and their queen, this behaviour can have a future.

The unlikely success of thermal scavenging is perhaps best illustrated by its independent evolution in three groups of ants living on distant continents. As well as *Cataglyphis* ants in Europe and North Africa, there are *Melophorus* ants in Australia and *Ocymyrmex* ants in the Namib Desert, neighbours of the fog-basking beetles. In 1916, George Arnold seemed delighted by the latter, writing in his otherwise very dry *A Monograph of the Formicidae of South Africa*, 'These ants are also endowed with the most marvellous celerity, far excelling in this respect all other ants with which I am acquainted, so much so that they appear almost to fly over the surface of the ground. Their motion is just as erratic as it is swift; they seem incapable of pursuing a straight path for more than a couple of inches, and watching one of these insects for but a short time, a casual observer might be led to suppose that the unfortunate animal, having lost its way, had been seized with hysterical dementia.'

Common to all thermophilic ants, this erratic sprinting isn't a sign of disorder, however. It, too, is an adaptation to extreme heat. Unlike ants that forage on leaves or the nests of termites, thermal scavengers could find food anywhere. It is unpredictable. The only search category they might use is that the next meal is likely to be in full sun. Even if a scavenger returns to a particularly bountiful area in which to forage, it has no way to communicate its location to its nestmates. Pheromone trails, the reason most ant species follow each other in a line, would quickly evaporate at the temperatures at which *Cataglyphis* ants forage. And so they are left with one option: scouring the dry earth by themselves, maniacally but methodically. If a carcass is too large for them to carry, they leave it. While they rub their antennae with hundreds of nestmates when underground, a foraging *Cataglyphis* ant is an individual that either finds food or dies alone.

During this late morning in mid-June, Cerdá bends over as he slowly meanders through patches of thistle looking for one species of ant in particular, a small member of the *Cataglyphis* clan and one of the most efficient thermal scavengers known. While *C. velox* is protected from the burning hot surface with long legs that lift its body off the ground by two millimetres, pushing it into air that is up to 15°C cooler, *C. rosenhaueri* has evolved an additional strategy. A specialized hinge between its thorax and abdomen – known as the gaster – allows this species to point its bum towards the sky as it runs, lifting it into the cooler air above. This simple trick, Cerdá has shown in the lab, allows these ants to cool their body by another 6°C (compared to just having long legs) when the ground is 55°C. After a few minutes of searching, Cerdá spots one and calls me over to his patch of litter-strewn earth. A speck no bigger than a poppy seed darts across the gravelly substrate, stopping frequently as if at an invisible crossroads. 'It is very, very fast,' he tells me. From above, it seems like the ant is shrinking as it runs, the upturned abdomen making the whole animal look off-balance or incomplete. In the erratic style that Arnold noted in its South African relatives, this ant might keep sprinting and searching for the next 40 minutes before returning to its nest. But the air temperature is already 30°C at our human-level, and I'm burning up in shorts and a T-shirt. We get back in Cerdá's car, feel the cool rush of air-conditioning and rejoin the motorway.

Now in his early fifties with short, curly white hair and a welcoming face creased with wrinkles, Cerdá has spent his life in pursuit of insects. As a boy, he collected anything that scuttled, jumped and flew around his home near Barcelona, northeastern Spain. With a net in hand, he was just another inquisitive child who found solace in the order of life, its species and families, and by the almost endless diversity in those animals that have six legs and three body segments – head, thorax, abdomen. Beetles with long horns, butterflies with forked tails, flies with an iridescent sheen. While his classmates were more interested in birds, Cerdá was happiest crawling on the ground looking for bugs. Most of the specimens he displayed in his room were very common and easy to catch, but he still thought they were special. Swallow-tail butterflies, named for their forked streamers on their hindwings, were his favourites.

Insect collecting is an almost limitless hobby, a fascination that seems to demand further study into adulthood. While there are 10,000 species of birds, the number of insect species is in the millions. Just considering beetles alone, those insects that have made their first pair of wings into a hard covering, or 'elytra', there are over 400,000 known species. In a potentially apocryphal attribution, the geneticist J.B.S. Haldane once said that if there was a Creator of life on Earth, he or she had 'an inordinate fondness for beetles'. By comparison, there are over 15,000 known species of ants, a group of insects nestled within the order Hymenoptera along with wasps and bees. But what ants may lack in diversity of shape and number of species, they make up for in sheer abundance. A study published in the journal *Proceedings of the National Academy of Sciences* in 2022, for example, estimated that there were 20 quadrillion (a million more than a trillion) ants living on Earth. If we collected them all up and handed them out as gifts to humankind, every person would have 2.5 million ants each. And if we put this ginormous bag of ants on some suitably ginormous scales, it would weigh more than all the wild mammals (including the titans: elephants, giraffes, whales) and birds put together. It would weigh two-thirds of *all* the insects.

But to bag up all the ants of the world would also be a devastating loss, depriving ecosystems on four continents of their most willing nutrient recyclers, the workers that collect organic matter – leaves, seeds, dead arthropods – and convert it into more ant-shaped biomass. E.O. Wilson wrote that ants are 'the little things that run the world'.

The study of ants is known as myrmecology. To study ants, therefore, is to be a myrmecologist. As we drive to his second field site outside of Seville, I ask whether Cerdá uses this label for himself, a badge of pride that connects him to the greats such as Wilson. He prefers entomologist, he tells me, a broader term for someone who studies insects in general. Indeed, after completing his PhD in ant anatomy, Cerdá studied the reproductive system of cockroaches, a very different branch of the insect family. This job would have led to a permanent position in his hometown of Barcelona. After two years of staring down microscopes and dissecting ovaries, however, he longed to be outside again, to observe insects in their natural environment like he did as a boy. For this, he accepted a position as an ecologist in Seville, where he now works at the Biological Research Station of Doñana, a modern, terracotta-clad building just over the river from the narrow cobbled streets, cathedrals and tapas bars of the city centre. Ants were never his primary motivation for research, or even his favourite insects. But what they gave him was a return to regular fieldwork; whether it's a patch of dirt next to a motorway in Seville or the sand dunes of the Moroccan desert, ants are everywhere.

Unfortunately for his three youngest children, Cerdá's field trips are often combined with family holidays. And when you study some of the most thermophilic animals known, that means a lot of time in the desert. His wife, Elena Angulo, also works at the Biological Research Station and also studies ants, particularly the invasive Argentine ant that is devastating local species around the world. In 2021, Cerdá and his family caught the ferry across the strait of Gibraltar and drove into southeastern Morocco, to the town of Merzouga that sits at the edge of the Sahara Desert. There are camels, quad bikes and the rolling sand dunes that seem barren of all life. Back at his office, Cerdá shows me a few photos from his trip, pointing out that even here there are regular patches of vegetation. And near these oases of greenery and shade, there are ants, such as *C. bombycina*, the Saharan silver ant. He shows me a photo of his son holding one of these ants in his hand, its silvery sheen almost as bright as the smile on his face. And then there's his youngest daughter, eight years old, having a nap in the shade of one of these bushes, sun hat pulled over her face. For at least two days during their trip, the kids can have unlimited access to the local waterpark.

The Saharan silver ant is perhaps the most famous – and not infamous – of all ants. It has featured in David Attenborough documentaries. It has graced the covers of scientific magazines. Undoubtedly it is because of its silvery coating, making each ant look like an animate blob of metal that scours the harshest habitats on Earth for food. No other *Catalglyphis* ants, even those that live nearby in Tunisian salt pans – the black *C. fortis* – have this. Their unique sheen allows this species to forage in temperatures up to 54°C,[1] making the Saharan silver ant one of the most thermo-tolerant animals on Earth.[2]

Their silver coat is actually the sun's rays reflecting off a fine covering of hairs. These aren't made of keratin but of chitin, a tough protein used by insects, other arthropods and fungi. Remove one of these horizontal hairs, cut across its width like chopping through a carrot, and it will look triangular under a microscope. This is what scientists did in a study published in *Science* in 2015. Nanfang Yu of Columbia University and his colleagues teamed up with Rüdiger Wehner to provide the most meticulous look at thermophilic ants to date. Under their high-powered microscope,[3] the side of these triangular hairs next to the ant's body is flat, following the contours of the head, thorax and abdomen. But the two sides that face upwards (and they always face upwards) are ridged, like the corrugated roof of a garage. This is crucial. It means that the sun's light is scattered in at least two different ways. Some of the light is reflected off the ridged exterior. Like throwing ping-pong balls on the garage roof, they bounce off in all directions. So does sunlight. But the sun's rays can also pass *into* the hollow hairs, deflecting slightly as they do (just as light is bent when it moves from air to water, known as refraction). This is where the flat bottom of the hairs becomes important: it acts as a mirror to these internal rays of light, bouncing them back upwards, away from the ant's body and back through the ridged surfaces where they are scattered into the surrounding environment.

Using fancy spectrometers, computer simulations and dipping into Kirchoff's law of thermal radiation (yeah, me neither), Yu and his colleagues weren't averse to the simpler methods of science. To reveal the importance of something, scientists often take it away. To understand what a gene does in development, it is inhibited or cut from the DNA sequence. To see the impact of a species on the wider ecosystem, they are excluded from a patch. To understand the cooling powers of *C. bombycina*, they shaved them. Without their triangular hairs, the heads of Saharan silver ants were no longer silver. And they reflected 41 per cent of the solar radiation that landed on their surface. With their hairs, however, 67 per cent was reflected. It's perhaps not a huge difference, but when you're walking on the thermal tightrope[4] that ants often find themselves on, 26 per cent is a huge window of opportunity, the difference between finding the next meal and *becoming* the next meal.

These experiments were on dead, often decapitated, ants. But desert ants don't just stand still and allow the sun's rays to slowly heat them up. When outside of their nest, they are nearly always on the move. As I saw with *C. velox* and *C. rosenhaueri* in southern Spain, they dart across their barren neighbourhood, their long legs making them the Usain Bolts of the bug world. In 2019, Sarah Pfeffer, an entomologist based in Ulm, Germany, and her colleagues pointed a high-speed camera at individual Saharan silver ants and found that they could run at 100 body lengths per second, nearly a metre per second for an animal the size of a sesame seed.

And sprint is the correct word here. They have to take regular breaks, finding thermal refuges that allow them to dissipate more heat when they are reaching their thermal limit. While the ground might be nearly 60°C, just a centimetre above can be 20°C cooler. A twig or a tuft of grass can be a lifesaver, allowing the ant to move into the cooler air above. The hotter it gets, the more an ant needs to stop at a thermal refuge. One study found these ants spend between 25 to 75 per cent of a foraging trip in these refuges.

Although they can survive in temperatures well below 30°C, desert ants avoid them largely due to competition with other ants. '*Aphaenogaster* is the main competitor of *Cataglyphis*,' Cerdá tells me, referring to the slow-moving, black ants with shorter legs that we saw earlier in the day. 'They are also scavengers.' But there's an important difference, he adds. Working in the slightly cooler hours of the day, *Aphaenogaster* can also utilize pheromone trails to guide their nestmates to larger prey that a solitary *Cataglyphis* couldn't carry. 'They can arrive with 20 workers or more to collect a big cricket,' he says, 'but *Cataglyphis*? No. *Cataglyphis* is [limited to] small pieces because they can't collect cooperatively. They are always solitary searchers, solitary transporters.'

Cataglyphis ants, like the crucian carp of Northern Europe and Asia, are known as subordinate members of their community. They don't fare well when other ants are around. They rarely fight. They still recognize their nestmates by their distinctive odour, palpitating each member with their antennae as they enter their nest. Peering into the entrances to a few nests, each one a volcano in miniature, I spot a few workers standing on guard like bouncers, admitting those workers returning to the nest with food. Subordinate they may be, but they will kill an ant from a neighbouring colony. Living in the hottest hours of the day, a time when food sources are unpredictable and infrequent, these ants still have a tendency towards violence when it comes to their neighbours.

Fortunately, an ant astray is a rarity. Desert ants have an extraordinary ability to know where they are, where they have been, and how to get back to where they were. They use a process known as path integration, and scientists from around the world study *Cataglyphis* ants to learn how such a small brain can achieve such complex tasks. As Charles Darwin noted in 1871, 'the brain of an ant is one of the most marvellous atoms of matter in the world, perhaps more so than the brain of a man'. The workers of *C. fortis*, for example, can roam 100 metres or more from their nest on the salt pans of Tunisia and then make a bee-line (ant-line?) back to where they started. While this trait isn't unique to these ants, it has been honed to a tee due to the inhospitable conditions in which they work. Sprinting through the hottest part of the day might reduce competition and predation, but it can't be maintained for long periods. Once a food item has been found, the ant has to get back quickly before their body overheats. To help, they memorize structures that may identify their nest, count the number of steps they take, and take note of the polarized light from the sun. Putting all these together gives them a coordinate. A desert ant might not know where their next meal is coming from but they have a very good grasp of where they are.

After they've exploded from their nest, desert ants have started an imaginary timer on their own survival. Sprinting on stilt-like legs allows them to survey a large area. Extraordinary resilience to heat and regular breaks keep them alive. Antennae taste the air and ground for signs of a dead insect. But all would be pointless if they didn't know how to get back to the colony. These ants, technically, don't thrive in the inhospitable. They move through it, almost killing themselves in order to survive. 'Thermophilic' is a misnomer, in other words; *Cataglyphis* ants don't love heat, they just deal with it better than other animals.

During my stay in Seville, as air temperatures in the city reached 44°C, I was behaving in a similar way to a desert ant. I wasn't sprinting over the cobbled streets and climbing up statues, but I was depending on thermal refuges to get by: shade, air-conditioned buildings, fans and cold drinks. I was using them all, sticking to the side of the street in the shade, finding a bench underneath an orange tree, refrigerating my bottle of water and staying inside during the hottest part of the day. With climate change, such refuges are set to become an essential part of life across much of Southern Europe.

While we may show similarities in behaviour, we can't directly compare the thermal maximums of ants and humans. We thermoregulate in very different ways. An ant is tiny and known as a poikilotherm, meaning that its body temperature can swing up and down with environmental temperatures. Humans, however, are homeotherms. We like to keep things stable, not drifting far from our ideal temperature of 37°C. Whether in the high Arctic or the African savanna, we maintain this body temperature otherwise we get ill and die. In the cold, we wrap up warm and we shiver. In the heat, as my shirts and boxers can attest from our trip to Seville, we sweat. As this liquid evaporates from our skin, it pulls some of the heat away from our skin's surface, cooling it. While there are complexities and caveats to this simplistic explainer of homeothermy, this is the essential takeaway: to cool, we use water. And that water needs to leave the body. This means that it's not just the temperature that can be lethal, but also the ability of water to escape. In a dry setting, it can almost explode from our skin, finding a large diffusion gradient to move into. But in a humid setting, the water from sweat finds it difficult to move. It is surrounded by water molecules in the air, all jostling for space. And so, to determine the critical thermal maximum of our bodies, we have to use a thing called the wet bulb thermometer, essentially a classic mercury thermometer wrapped in a wet cloth. In dry air, the wet bulb will cool much easier than in humid air. The evaporative cooling will reduce the temperature, meaning that the mercury doesn't travel as far upwards.

Using this technique, scientists have found that the critical thermal maximum for humans is 35°C with 100 per cent humidity. Currently, such an environment is very rare. But projections of climate change estimate that vast regions of Central Asia, the Gulf and Northeast Africa will regularly experience such hot and humid air by 2050. Plus, this is only the upper limit. It isn't the temperature when our bodies suddenly succumb to heat stroke, a potentially fatal reaction that affects some people more than others. As a report from the Centers for Disease and Control and Prevention states, 'Outside workers, older adults, children, communities of color, the homeless, individuals with a mental health disability, individuals with chronic medical conditions, individuals without access to air-conditioning, and low-income communities are particularly vulnerable to heat-related illness.' Although it is difficult to know exactly how many people die every year from the heat (a heart attack might be caused by a number of health conditions),[5] one study estimated that the hot summer of 2022 led to over 60,000 deaths in Europe alone. Another study, published in July 2021, estimated that, globally, nearly half a million people died from heat stress every year between 2000 and 2019. While cold-related deaths were ten times as frequent (nearly five million people), they were also decreasing around the world. Heat-related deaths, however, were increasing year on year, particularly in 'large, low-lying, crowded coastal cities in Eastern and Southern Asia and cities in Eastern and Western Europe'.

Studying *Cataglyphis* ants might not provide any insights into our own physiological response to heat, but their lifestyle nevertheless comes with a lesson: even the most heat-tolerant insects have to find ways to cool down.

In contrast to extreme cold that constricts biological machinery and cell membranes, heat leads to fluidity, a vibration of kinetic energy. After all, heat is ultimately the movement of atoms, a nanoscopic shuffling that we detect as warmth or fierce heat. This has been known for a long time, even before atoms were discovered. 'Heat,' John Locke, an English philosopher of the Enlightenment Period, wrote in 1720, 'is a very brisk agitation of the insensible parts of the object, which produces in us that sensation from whence we denominate the object hot; so what in our sensation is heat, in the object is nothing but motion.'[6]

As a prime example of this heat-as-motion theory Locke would surely have been delighted to know about the Japanese honeybee. To defend their nests against their main predator, the five-centimetre-long Asian hornet, hundreds of honeybees emerge from their colony and surround the intruder in a ball of buzzing insect bodies. Originally thought to be an overzealous suicide mission – those bees at the centre of the ball stinging the hornet to death – the reality is far more subtle. By vibrating their wing muscles, the honeybees generate enough heat to increase the temperature at the centre of the ball to around 46°C. Half an hour after this bee ball first forms, the hornet has been pushed beyond its thermal limit. It hasn't cooked, as is often written, but it has been killed. Importantly, the honeybee can endure temperatures up to 48°C. The difference between life and death can be a mere couple of degrees.

What can explain this slight difference between a honeybee and a hornet? The thermal limit of *Cataglyphis* and *Aphaenogaster* ants? Why does the lethal temperature of any species differ? A large part of it boils down to a group of proteins suitably called heat shock proteins, or Hsps for short. First discovered in the 1960s, Hsps have now been found in every organism, from microbes to humans, and represent a universal response to thermal stress. An Arctic fish that is most comfortable in 0 °C water will produce these proteins at just 5 °C. For a fruit fly, significant thermal stress begins at 33 °C and stretches up to 37 °C, temperatures at which Hsps are at their maximum rate of production.[7] All other proteins that the cell usually produces are, for a time, halted entirely. Extreme heat, even at a subcellular level, is overwhelming. When pushed beyond a certain comfort zone, these proteins can hold things together, at least for a time. The increase in motion can contort proteins into dysfunctional shapes, preventing important cellular processes like DNA replication. Heat shock proteins, however, can provide support as the thermal jostling increases, a cushion against chaos. While antifreeze proteins bind to water molecules to prevent ice crystal formation, Hsps bind to the biological machinery inside a cell, wrapping it in a protective case. Just as a plaster cast holds a broken bone in place, some heat shock proteins (there are hundreds of different types) can even reform those proteins that have been broken, returning them to their 'native state'.

While it was once thought that Hsps were an emergency response to overcome brief moments of stress, *Cataglyphis* ants were some of the first animals to show that these proteins can be an everyday part of an animal's being. The Saharan silver ant, for example, will produce the same amount of Hsps at 25 °C as at 50 °C. Unlike other species of insects, they don't seem to have a heat shock *response*, in other words. They are continually prepared for heat. This might be because every day is a shock to them, emerging from their nests with body temperatures less than 30 °C and then, within a few seconds, rising to 50 °C or more. Rather than donning their armour at the battlefield, these ants are fully clad inside their nests, just waiting for the right moment to explode into the day's onslaught of heat.

Heat shock proteins are even more important than their name suggests. As the delightful 2014 textbook *Heat Shock Proteins and Whole Body Adaptation to Extreme Environments* states, they should really be called 'stress proteins' since they are also protective against hypoxia, toxic metals, pesticides, UV radiation, viral infection, desiccation, salinity and other environmental insults. In recent years, they have even been associated with neurodegenerative disorders such as Alzheimer's disease. Also known as protein misfolding disorders, the build-up of tau proteins and brain damage that define Alzheimer's have been linked to a particular type of heat shock protein called Hsp70. In mouse models of Alzheimer's, an abundance of another heat shock protein, Hsp27, can prevent the symptoms of disease altogether. An intranasal spray that can deliver Hsp27 to a mouse's brain has shown ameliorative effects in these mice, reducing the death of brain cells and improving memory. First discovered in fruit flies that were warmed outside their comfort zone, Hsps may offer a new way of treating one of the most common and debilitating diseases that is predicted to afflict 130 million people around the world by 2050.

The importance of heat shock proteins in health and disease makes sense from an evolutionary point of view. From its origin, life has been moulded by heat.

There are three leading theories as to where life emerged on Earth, and all three are hydrothermal – very warm to hot in temperature. While the geological record of this moment is lost to billions of years of planetary change, ancestry is also written into the DNA of microbes alive today. Genetic sequencing is a telescope into the distant past. By analysing the DNA of species from several branches on the tree of life, scientists can extrapolate back to how their ancestors may have lived. Looking at the oldest members of bacteria and archaea, for example, they are replete with heat-loving species. The roots of their evolutionary tree are embedded in hot, often boiling, water. The so-called hyperthermophile Eden hypothesis is supported by the DNA of microbes alive today.[8] While the precise location of the event is unknown, scientists who study life's origins generally think of it as being hot.

When life emerged is also unknown, blurred by the passing of four billion years. The oldest fossils suggest that habitable conditions began around 3.8 billion years ago, a time that followed the 'Great Bombardment' that saw the Earth pummelled by asteroids and other space rocks. The largest of these happened when the planet was first coalescing into a mass of spherical stardust. A roving planetary body the size of Mars collided with this proto-Earth; a golf ball piercing through a water balloon. Informally called the 'Big Splat', any water was instantly boiled and ejected into space. The impact produced so much heat that even rock vaporized, a solid converted straight into a gas. While some of this rock cooled and formed the moon, the majority created a cloud of dust that blocked out the rays of the sun. Earth was a planet of magma oceans where air temperatures rose to 2,000°C, nearly half the heat of the sun's surface. For thousands of years, Earth was a blur that radiated its violent birth into the atmosphere and quickly – geologically speaking – cooled. After 20 million years or so, the rock solidified into a new crust: bedrock, a foundation. The atmosphere was rich in carbon dioxide and low in oxygen. There were only small islands of land poking through a near-continuous ocean. This planet was so different to modern Earth that it may as well have a different name.

In the aftermath of this interplanetary violence, chemistry became contained and started to replicate. Life emerged. While the simplest explanation is that life emerged just once, a microbe that would inoculate the world with wonder, there are scientists who think that life emerged whenever it could. There is an inevitability to biology. The most brazen of these researchers have even proposed that microbial life may have emerged in the superheated, molten chaos that we call the Hadean Period (over four billion years ago) and survived even the greatest bombardments. While there is no evidence of such a theory in the geological record, it shows their unwavering respect for life's resilience.

The study of heat-tolerant microbes helped create the study of extremophiles in the first place, an early acknowledgement that life could thrive well outside of our comfort zone. It was in the 1960s, and Thomas Brock, a young microbiologist from the University of Illinois, travelled to the hot springs of Wyoming's Yellowstone National Park, a place where volcanic activity under the surface superheats water to create spectacular plumes of vapour, known as geysers, and boils mineral-rich puddles and ponds. Where these minerals settle, colourful bands of turquoise, tangerine orange and mustard yellow encircle the hydrothermal centres, like a geological vortex slowly growing around the eye of a storm. Dressed in brown trousers, comfortable shoes, a flat cap, and carrying a leather satchel, Brock dipped 30 microscope slides – rectangles of thin glass the length of a little finger – into water that was scalding hot. Then he transported his samples back to the lab and peered at them through his microscope. He discovered a new species of bacteria, the first hyperthermophile, an organism that prefers temperatures above 80 °C.

Just a few months before this, an influential paper in the esteemed journal *Science* claimed that what Brock had just observed was an impossibility. Published in December 1963, its author Ellis Kempner, a researcher at the National Institute of Arthritis and Metabolic Diseases, wrote that the hottest temperature that life – microbial or otherwise – can withstand is 73 °C. Like Brock, Kempner had also sampled some water from Yellowstone's hot springs and found that no microbe could withstand temperatures above this. Previous authors that had claimed to have found life in 89 °C, he added, were mistaken. At such temperatures, DNA would separate. Amino acids, the building blocks of proteins, wouldn't line up into the proper sequence. RNA, the molecules that translate DNA code into recipes for proteins, would degrade. 'Whatever its molecular basis,' Kempner wrote, 'it is clear that there is a maximum temperature for active life processes. The earlier ecological reports which have been widely quoted must therefore be reinterpreted as survival without metabolism.'

Survival without metabolism… Meaning: everything above 73 °C is in a dormant state waiting for conditions to improve, an organism closer to death than life. But as confident as his claims appear, Kempner should probably have stuck to arthritis and metabolic diseases. Just a few years later, Brock published his findings from Yellowstone's hot springs in the less well-renowned *Journal of Bacteriology*. The slime on his glass slides not only survived in 73 °C, they thrived. They also grew at 74, 75 and 76 °C, up to a limit (in the laboratory) of 79 °C. Survival with metabolism. Growth. The thermal limit to life had been broken.

The current record for the hottest hyperthermophile is 121 °C. Some scientists think that the actual limit may be closer to 150 °C.

Along with his ironically named student Hudson Freeze, Brock named the bacteria that coated his slides *Thermus aquaticus*, a microbe that lives in water (*aquaticus*) and likes it hot (*Thermus*). While it was extraordinary in terms of its niche, it was a remarkably boring species in every other respect. As Brock and Freeze wrote in 1969, 'This organism is a gram-negative, nonmotile, nonsporulating rod'. It didn't stain, it didn't move, and it didn't create spores. A species defined by absence. Even in the world of bacteria, it was a pretty bland specimen, the vanilla of microbial fauna.

And yet, this plain-looking bacteria would be the seed that germinated a whole field of biology, the study of extremophiles. Organisms that love extremes. Whenever such organisms are reviewed in the scientific literature, the references almost always point back to 1969, to the *Journal of Bacteriology*, to Thomas Brock. More than this, *T. aquaticus* has revolutionized pretty much every part of biology, becoming such an integral part of the modern laboratory that few realize just how important this microbe is. When I was a biology student studying the genetic diversity of fish from sharks to ocean sunfish, I used it all the time without even noticing. It is called the polymerase chain reaction. You might recognize its initialism, PCR. From COVID tests to forensic analysis, *T. aquaticus* has been indispensable. The backstory to PCR is long and winding, and involves LSD. For now, let's just say that this laboratory technique can produce a lot of DNA from a very small amount of DNA. It also requires heat. And an enzyme taken from *T. aquaticus* was able to withstand this heat without – as many other enzymes would – becoming unravelled.

With this thermo-tolerant ingredient, PCR could be automated with a machine. Billions of strands of DNA could be generated from a speck of blood or a drop of mucus in an afternoon. 'PCR revolutionized everything,' David Bilder, a professor at University of California, Berkeley, once wrote. 'It really superpowered molecular biology – which then transformed other fields, even distant ones like ecology and evolution… The ability to generate as much DNA of a specific sequence as you want, starting from a few simple chemicals and some temperature changes – it's just magical.' That magic came from evolution's command over heat.

We may think of a superheated world as a planet on fire. But the Earth didn't flicker in flame until much later in its history. Fire can only emerge in the presence of oxygen and a flammable – often carbon-rich – material. Itself needing to breathe and feed, fire is a by-product of a living planet. By 420 million years ago, a time when life was making its first forays onto land, the amount of oxygen in the atmosphere had reached levels at which heat could turn to fire, and fire could stay lit. By the Carboniferous Period, 359 to 299 million years ago, not only was the percentage of oxygen in the air around 30 per cent, nearly a third higher than today, there was a wealth of combustible material: the carbon-rich, woody tissue of the first trees. Growing to 20 metres tall or more, ancestors of horsetails, a group of fir-like plants that gardeners consider to be weeds, formed dense thickets in the muggy swamps of this period. With so much oxygen in the air, however, even damp wood could ignite from a lightning strike. As *Meganeura* dragonflies the size of crows tormented the air,[9] these early forests burned with only the slightest persuasion.

Since its first flicker all those millions of years ago, fire has been a dominant force on our planet, a pressure that life has to adapt to in order to survive. Most animals can avoid being cooked, either by burying into the earth, hiding in a hollow trunk or fleeing an area altogether. (Some species of predatory birds actually pick up burning sticks and drop them on new areas in order to flush out their prey.) But plants are not like this. Plants can't move, never mind flee. They are rooted to the Earth and, when conditions dry out, they are also highly flammable. Across the Mediterranean-like ecosystems in Europe, the Americas and Australia, regular fire seasons have fashioned a suite of remarkable adaptations not only to endure the heat of a wildfire but to depend on it, just as another plant might depend on light and water.

There are two broad categories of fire adaptation. The first is to make sure you're ready to regrow once the fire has passed, an ecosystem ripe for the taking and fertilized by the ash of dead neighbours. Spending years or decades building up a seed bank in the soil is perfect for this. But there has long been an active debate on whether these traits, such as growing from buds under the ground or beneath a layer of bark (post-fire regeneration), or from seeds in the soil (post-fire recruitment), are simply responses to disturbance of any kind. Growing quickly following a drought, flood or consumption by herds of animals or swarms of insects, for example. Disturbance, rather than fire adaptation, is a trait common to a lot of plants. If they weren't primed to respond to a change in their environment, they would soon be outcompeted.

So, to avoid the heated fury of fire ecologists, let's focus on those plants that are – almost certainly – dependent on fire. One of the most common and best-studied groups of plants is the humblest of conifers: pine trees. Planted and nurtured for their timber, brought indoors at Christmas time, the fire adaptations of these trees have their origins deep in the Cretaceous Period, roughly 135 million years ago, a time when dinosaurs were getting extremely big, pterosaurs soared through the skies above and shrew-like mammals, our ancestors, were small and often hiding in subterranean burrows. It was also a very flammable time to be alive. Oxygen levels were over 26 per cent, 5 per cent higher than today. And the climate was often warm and dry. This combination, one study notes, likely made the Cretaceous 'one of the most flammable periods in the Earth's history'.

Our own evolution was dependent, indirectly, on the fires of the Cretaceous. The ripe fruits and leaves that primates would later select with colour vision and dexterous fingers, all growing on the group of plants known as angiosperms, took root during this time in Earth's history. Growing in the shadow of the conifer forests, these fruit-bearing plants were restricted to the 'dark, damp and disturbed' places on land. Just as mammals were sheltering under the feet of dinosaurs, the early angiosperms were the subordinates of plantdom, eking out a living under the buttresses and canopy of giants. Fire in the Cretaceous may have changed this balance of power away from the pines and cycads and towards the members of the understorey that could grow quickly from well-nourished seeds. As just another disturbance, the angiosperms thrived in the fire's wake. Fossil evidence shows that by 100 million years ago, a time when oxygen levels were above 26 per cent and fires would have been commonplace, the angiosperms were diversifying, creeping out of the shadows. Today, they are the dominant form of plants in most ecosystems, from tropical rainforests to the flowers in your garden. Given the opportunity to emerge from the shadows, they grew the fruit that our ancestors plucked and ate, powering the brains that would one day become ours.

The burning presence of fire is recorded in the cold rock of the fossil record. First, there is charcoal, the blackened remains of woody plants that burned long ago. But there's also the signs of fire as shown in how plants change over millions of years. In the Cretaceous, for example, relatives of modern pine trees started to wrap their trunks in a thick covering of bark, a few millimetres expanding to over three centimetres. It might not sound like much, but this was a shield against some of the hottest and most enduring fires. A 15 millimetre layer of bark, for example, will protect a tree's delicate cambium, the sugary phloem and watery xylem at its core, for a few minutes of a low-intensity surface fire, the type that stays low to the ground and reaches temperatures of 400 °C. Double this thickness, however, and a tree can survive a crown fire, the acme of flammability that reaches into the canopy, is stoked by the wind and can reach 800 °C. Pines that grew bark over three centimetres thick could survive this searing heat for over ten minutes.

Many pines welcomed crown fires just as farmers welcome rain after long periods of drought. Looking at the fossilized pine cones from this time, as well as those of pines alive today, one of their key traits is a thick, sticky resin, a type of botanical glue that only melts at extremely high temperatures. With their cones held aloft in their higher branches, these giants of the forest would only reproduce when a fire erupted into their canopy. With the resin melted, the scales of the cones would open and their winged seeds be blown by the wind, settling in an area with reduced competition, increased light and a thick layer of nutrient-rich ash. A type of serotiny (meaning 'following,' or 'later'), this is a feature that allowed pines to thrive in the Cretaceous and reach into the modern day. Lodgepole pines in the western United States are classic examples of fire-related serotiny.

Dylan Schwilk, an ecologist who studies fire adaptations, shows me the lump of sequoia tree he keeps in his office. It's primarily made up of bark with a small circle of wood at its centre. Giant sequoias, a species of conifer belonging to the same family as cypress trees, can grow to over eight metres in diameter, a lighthouse-sized life form. Over a metre of this can be bark, dead wood, a shield against the fiercest of fires. This particular specimen was used as a sign in an agricultural show, and then discarded. Asking a dendrochronologist to take a look at the tree growth rings, Schwilk was told that this tree was chopped down in 1882. As sequoias can live for over 2,000 years, its burst from seed following a fire may have happened when the Romans were conquering Europe. Whenever the true birth date was, this tree would have germinated in a world that was very different to that when it was cut down. North America would've been home to tribes of Native Americans, people who regularly burned forests in order to regenerate the flora on which they depended. While not technically serotinous, the seeds of giant sequoias germinate best in the ashy, mineral-rich soils that are left in the fire's wake. ('They are about as dependent on fire for successful recruitment as is a serotinous pine,' Schwilk told me in a follow-up email.) With the arrival of Europeans, however, these traditions were lost. The Native Americans were slaughtered, died from Western diseases, and thousands of years of culture was reduced to whispers. 'Fire was seen as an unnatural force,' Schwilk says. 'It was based on the European view of deciduous forests.'

Without fires set by the Native Americans, the redwood forests of the western United States didn't set seed. There was no recruitment. For 100 years, no fire meant no offspring. It has been called the 'lost generation' of redwoods. 'Of course, for a plant that lives for 2,000 years, it's not a big deal,' Schwilk says. But it shows the importance of fire in ecosystems that were originally forged in flame. No fire, no offspring. An ancient forest without a future.

Realizing the importance of fire to giant sequoias, the National Park Service began conducting prescribed burns in California's Sequoia National Park in the 1960s, a practice that Native American tribes are now returning to. By cutting back, clearing and burning the understorey that has accumulated beneath these ancient giants, a historical symbiosis is being brought back to life. 'There's a lot of cultural burning practice and projects happening now,' Beth Rose Middleton Manning, a professor of Native American studies at the University of California, Davis, told *The New York Times* in 2024. 'It's helping improve forest resilience, especially in the face of climate change.'

Pine trees weren't mere bystanders in their 135-million-year-old story. As their bark thickened and their cones became stiff with wax, the Cretaceous pines also started to retain their dead branches as they grew. This old wood formed a dry and highly flammable skirt. Unsightly to the modern forester or hiker, a tangle of death hanging towards the ground was, and still is, a crucial part of their adaptation to fire. In what fire ecologists have called 'niche construction', these dead branches act like kindling to any fire that spreads over the surface of the ground. With branch retention, the theory goes, a low-intensity fire could be persuaded to grow upwards into a full crown fire. A 400 °C burn becomes an 800 °C furnace. No longer holding water, sugars or photosynthesizing leaves, these dead branches are essential to the life of a fire-dependent pine. Without them, the fire may remain at ground level, the cones in the canopy closed, the seeds still glued inside. 'Fire-spawned stands of lodgepole pine are, in a sense, locked into a fire cycle,' David Pitt-Brooke wrote in his book *Crossing Home Ground*. 'They are creations of fire and, in turn, they create conditions hospitable to future fire. You could almost think of it as a symbiosis.'

Branch retention isn't unique to pines. In the summer of 1999, Schwilk investigated the dead branches of chamise (*Adenostoma fasciculatum*), a flowering shrub that is characteristic of the dry, California chaparral. Ducking under the one- or two-metre-high canopy, Schwilk started to modify a few plots of chamise that were each the area of a studio apartment. In some plots, he removed the dead branches and in others he left them alone. This simple experimental setup then allowed him to see whether even these small shrubs could alter their flammability. Then he waited for the prescribed burns that were planned by an ongoing study at the University of California, Berkeley, set for November that same year, a time of extreme aridity following the summer drought. In each plot, Schwilk had placed strips of copper painted with bands of heat-sensitive paint. Changing texture at different temperatures, 100°C, 150°C, 200°C, all the way up to 850°C, these paints would act as thermometers that could be read long after the fire had died and the plots cooled down. Another measure was a two-litre can filled with water, a small hole punched into the top. The amount of heat from the fire would cause steam to be released from the can. The amount of water left after the fire could be used as a measure of its intensity.

His results were stark. Out of 29 plots that burned, those that had their dead branches removed experienced a fire intensity of 150°C. Those that were 'un-manipulated' were over 100°C hotter, reaching nearly 300°C in places.

Perhaps this is unsurprising. The dead branches that were retained could simply have been more fuel for the fire to burn. But to test whether this was the case, Schwilk included a 'Clip and Leave' series of plots in which he, as you may have guessed, cut the dead branches from chamise but left them on the ground. They didn't burn any hotter than the 'Removal' plots. If anything, they were a little cooler (if 130°C can be called cool). It wasn't just the fuel load, in other words. The dead branches had to be *retained*. As anyone who has ever made a fire knows, the best fires aren't always made from a mass of firewood. It's how the sticks and logs are arranged that's important.

Chamise exemplifies a plant community that spans the globe: the Mediterranean-type flora that stretches from the Americas, through Europe and Asia, and down to Australia. While only 5 per cent of Earth's total land area, it accounts for 20 per cent of known plant species. Partly, this is because these areas are disturbed – by fire and drought, by hot summers and wet winters. An unstable habitat swings between hospitable and inhospitable, promoting an ever-changing cast of characters.

Disturbance is a delicate balance, however. It can easily swing into a depauperate system where only the most hardy of plants – we might call them weeds – dominate. This is what has happened for much of the California chaparral. Human-caused fires, whether accidental or prescribed, have pushed the shrubland into a wasteland dominated by fast-growing and fast-dying weeds, a gravelly bed of annual plants. Even fire-adapted plants need long periods without fire. What was once a 30-year event, caused by lightning strikes, is now happening every few years. The flammable chamise can't grow fast enough to set seed before the next fire. Similarly, a study into lodgepole pines from 2019 found that frequent burns are slowly reducing the ability of these trees to recover and reseed, 'initiating a downward ratchet until trees do not regenerate'.

This is a story that has engulfed the world. As climate change and human actions ignite fires that are hotter and longer-lasting than ecosystems have adapted to, plants are struggling to keep up. Even ecosystems that have never experienced fire have burned in recent years. Considering the Scottish fires of 2023, Canada's Fort McMurray fires of 2016, or the Australian bushfires that have grown in intensity year on year, Earth might be swinging back to a period when fire was the norm. 'Should the feedback loop of heating and drying continue to intensify as it has been,' John Vaillant writes in his book *Fire Weather*, 'there is in our future a potentially winter-less scenario in which fire weather is the only weather, and "fire season" never ends.'

There is a red weather alert sent out to all researchers at the Doñana Biological Research Station on the day I head south towards the coast. As temperatures are set to reach into the mid-40s by afternoon, all employees (and their visitors) are advised to be indoors, cooled by air-conditioning, by 12pm. With this in mind, I'm up before 5am to prepare for the day and walk through town to meet Jorge Isla, a tall and taciturn research assistant who will be my guide for the day. While I remember to set my alarm, I forget to charge my phone and I get lost in the cobbled streets that, at this hour, look completely different to the bustling bars and tapas bars the night before. To get to the Biological Research Station on time, I use a tourist map to guide me, knowing that it is over the Guadalquivir River and next to a theme park. I think of the desert ants that can guide themselves by the sun's rays and their memory of steps taken. I take the right roads but in the wrong direction, and I begin to jog to make it on time. When I arrive, I'm sweaty and embarrassed, feeling an even deeper appreciation for the ants I aim to see that day.

As with Mordor, one does not simply walk into Doñana National Park. It is fenced with thick security gates that control all visitors. As it is now a political hotspot due to the debate over strawberry farmers and their use of local water that once flowed into this wetland ecosystem, I have been told not to take photographs. Isla, however, drives the hour or so down the park several times a week and takes a lot of photos. After deflating the tyres of his 4×4 vehicle so we can drive on sandy tracks, he is carrying out one of his routine tasks: changing the batteries on dozens of remote cameras, collecting their SD cards full of photos. His supervisor studies the movement and activity of grazers – horses, cows, fallow deer, wild boar – and how it affects the marshland vegetation during drought.

Hot take: not well. The constant cropping of new shoots and shrubs exposes the ground to the full heat of the sun, baking the soil and turning the grass and reeds into a short stubble.

Vegetation is a stretch; marshland, a memory. As we drive through the park, Isla points out the places where water used to be and where it now sits: a few shallow ponds where we see a few flamingos and four spoonbills, all sifting the water for their microscopic prey. The dried stalks of grass show no sign of regrowth and the few cork oak trees that grow in this forgotten flood plain look so brittle that they would crumble into ash if I touched them. Isla tells me that rainfall hasn't changed significantly over the years. 'But the temperature is so high that everything dries.' Exacerbating this heat are a number of other stressors: the grazers, the strawberry plantations that use 85 per cent of the water that flows into Doñana in the spring, and the beach resorts that need drinking water and showers in the summer. All this affects the productivity of these habitats that were once flooded with river water and flushed by tidal seawater. But since the 1950s, the river has been shunted onto a new course for irrigating crops and the tide has been blocked with sea walls.

In 2022, the number of birds in the park was at its lowest since these populations were first counted in the 1980s. Insect populations are also falling. And all this has knock-on effects, especially for scavengers that depend on a rich biomass cycle of life and death to nourish their young.

By 11:42, the air temperature is 33°C and the sand is between 47°C and 51°C, depending on whether it is light or dark in colour. The *Aphaenogaster* nests are closed, their entrances blocked with a fresh covering of soil. Then I see the characteristic flash of an insect, a sprinting ant with long legs and a black body: *Cataglyphis velox*. As my eyes adjust, I see that they are everywhere on this patch of sand dune covered in prickly scrub. One is carrying a seed that is twice its size or more. The others are still empty handed (legged?), scouring over the hot sand in search of their next carcass. I have been up for hours by this point, but their day is only just beginning. They've cleaned their nests and will spend the next few hours in an inhospitable niche that opens and closes with temperature.

I want to stay and look for more ants but the temperature is now in the mid-30s and Isla has just finished pumping up the tyres for the drive back to Seville. It is also nearly 12pm, and the red weather warning is at the front of both of our minds. I didn't see the *C. rosenhaueri*, but perhaps it isn't hot enough for their thermal tightrope. This species, Xim Cerdá has found, tends to appear a few degrees after *C. velox*. This red weather that we are avoiding is their home. As we drive towards the motorway, flashes of bright blue bee-eaters erupting from the roadside, I'm constantly trying to locate areas of *Cataglyphis* ants: patches of dirt strewn with rubbish or sand dunes encrusted with dried grass. But they don't so much exist in a place as inhabit a period of time. And that time is now.

PART THREE

Rays

CHAPTER SEVEN

AIN'T NO SUNSHINE

Darkness

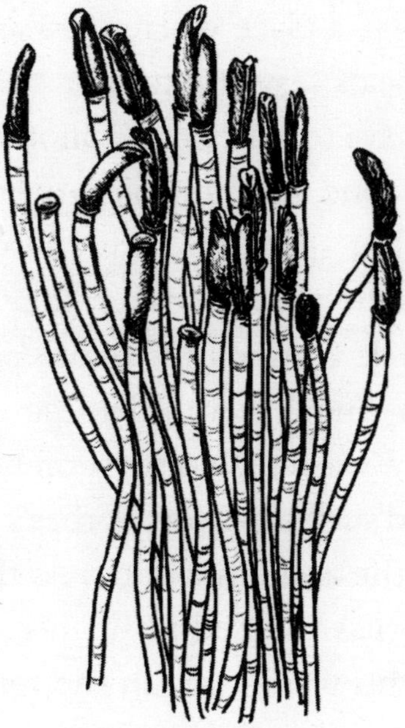

Attached to a four-kilometre-long wire, ANGUS was a two-tonne behemoth of sensory equipment that was suspended above the seafloor near the Galápagos Islands. On the research vessel above, a group of geophysicists, geologists and chemists were gathered around their computer screens, watching the information ANGUS was relaying back to them. Some were monitoring the depth, avoiding crash landings or collisions. Others were keeping an eye on the temperature. Just as the world above approached midnight, ANGUS sent back a spike in temperature. It lasted for three minutes and then disappeared. The camera on board was taking photos throughout but they would need to be developed once the submersible had returned to the surface. The year was 1977. What created this thermal anomaly? If the researchers were correct, this was a hot spring of the deep, a place where the Earth's mantle separates and superheats seawater to over 400°C.

They didn't have to wait for the photos to be developed. There was a second submersible, *Alvin*, that didn't just have a camera but space for three humans on board as well. At sunrise the next morning, they descended. Dive number 713. Jack Donnelly was the pilot. Jack Corliss and Tjeerd van Andel were scientific observers, geologists both. There wasn't a biologist on board. They didn't think they'd need such expertise; Forbes's Azoic Hypothesis still lingered in the academic mind. As they reached their destination, Corliss radioed up to his graduate student Debra Stakes who was sitting in the research vessel two kilometres above.

'Isn't the deep ocean supposed to be like a desert?'

'Yes,' she replied.

'Well, there's all these animals down here.'

One of these 'hydrothermal vent' ecosystems would later be called the Garden of Eden. Another, the Rose Garden. Swaying in the ocean currents, there were creatures with flower-like growths erupting from their pale and cylindrical bodies. Some were two metres long, a towering tube longer than any of the three men tucked inside *Alvin*. But they weren't roses. They weren't even plants. They were animals: giant tube worms belonging to the Pogonophora family of marine annelids (segmented worms). Anchored to the chimneys of these hydrothermal vents, they were encased in a flexible white shell, a protective cocoon into which they could retract their feathery plumes. Assessing 63 worms collected from the Galápagos Rift by *Alvin*, ranging in size from a few millimetres to over a metre and a half, the invertebrate zoologist Meredith Jones provided the specimen description and name, *Riftia pachyptila*, published in *Science* in July 1981.

A moustachioed specialist in worms, Jones was stunned by what he found. Or, more accurately, what he didn't find. As he teased apart the tissues of a 1.5-metre-long worm, there was no digestive system. No mouth. No anus. As one science writer would later say, it had 'no way in, and no way out'. How did this giant worm feed? Simple creatures can sustain themselves by diffusion across their membranes. Sponges don't have mouths. Perhaps *R. pachyptila* was doing just that, absorbing nutrients from its surroundings like an amoeba. Yet however simple its tissues, a giant tube worm is, well, giant. The same width as a rolling pin, it was a vast distance for nutrients to flow through. To answer this conundrum, Jones turned to the port-red protuberances at the end of the worm. Its featheriness was due to hundreds of tentacles attached to hundreds of grooves. He estimated that every worm had over 200,000 tiny tentacles wafting through the seawater surrounding the hydrothermal vent. It was a classic example of increasing the surface area for absorption of nutrients. Just as our own digestive systems are covered with tiny projections called villi that, in turn, are dotted with microvilli, these red tentacles would be in contact with a huge volume of seawater, and any nutrients it contains. This was how they supported their gigantism, he thought. Published in his 1981 paper, it was a solid hypothesis. It made sense.

It was also wrong.

A graduate student at Harvard University solved the mystery of how these worms nourished their bodies. Colleen Cavanaugh sat listening to one of Jones's lectures on his giant tube worms. He mentioned that he had found sulphur crystals in its innards, in an organ known as the trophosome. This cavity inside the tube worms made up a large amount of their body's volume, and Jones, rightly, assumed it had an important job. He surmised that the sulphur crystals it contained were the end-product of a detoxification process. Hydrothermal vents spew a lot of gases that can be toxic to animal life, and one of them is hydrogen sulphide. To counter this, did the trophosome contain a chemical toolkit that could detoxify this gas, turning it into an inert, harmless crystal? 'It's clear!' Cavanaugh interrupted, standing from her seat and snapping her peers to attention. 'They must have symbiotic sulphur-oxidizing bacteria inside of their tissues. They are feeding the worm!' The sulphur crystals, she had deduced, were the waste products of microbes that *feed* on hydrogen sulphide and release products that the worms can use to power their own cells. While toxic to animals, there were plentiful examples of microbes that used hydrogen sulphide to fuel their metabolisms. At this time in the early 1980s, the concept of mutualism – markedly different species living together for mutual gain – that Lynn Margulis championed in her 'endo-symbiosis theory' was revolutionizing the study of evolution. Cavanaugh had just found one of life's most spectacular symbioses that had been hiding in the deep.

In 1981, while still a graduate student at Harvard, Cavanaugh published her first paper into her theory, working with Meredith Jones as a co-author. Using an electron microscope that can detect the smallest bacterial cells, she found that 'characteristic lobes of the trophosomal tissue consist of densely packed spherical bodies'. Based on their shape and cell wall, the scientists were pretty sure they were bacterial cells. Importantly, they were 'far in excess of what could be attributed to bacterial contamination', a glimpse into the reality of the worm's innards rather than the by-product of a laboratory mishap. Later studies would unpick the finer details of this relationship. The feathery plumes of the worm are essentially a gill, a large surface that absorbs gases from the water. In this case, oxygen, carbon dioxide and hydrogen sulphide are funnelled through the bloodstream to the trophosome and its bacterial inhabitants. Using the sulphur and oxygen to power their metabolism, they release sugars that are formed from the carbon held in CO_2. As payment for their safe housing, the host worm gets some of these products. A two-metre-long worm is nourished on the scraps of billions of bacteria.

No mouth. No anus. No problem.

As well as imaging their trophosome with high-powered microscopes, Cavanaugh travelled down to see giant tube worms in the flesh, face-to-plume, in the ever-reliable *Alvin* (the same submersible that Mackenzie Gerringer travelled in, testing its latest limit of 6,500 metres). Later described as a character fit for a Jules Verne novel, a 'scientific Captain Nemo', Cavanaugh's work would help flip our view of life on its head. For centuries, the basic lesson in biology was that life needs sunlight for photosynthesis, a process of converting water and carbon dioxide into sugars. It was the original fuel of every food web. Even in darkness, nutrients would still filter down into ecosystems from this sunlit source. But here was an ecosystem completely ignorant of this zoological dogma.[1] The hydrothermal vent biota was sustained entirely by bacteria that eat hydrogen sulphide. Growing in a carpet over the surface of rocks as well as inside the animals that adapted to these vent ecosystems, these were the primary producers – the 'plants' – of a secret way of life. Each hydrothermal vent ecosystem was a world built by *chemo*synthesis.

Since the famous descent in 1977, dozens of hydrothermal vent ecosystems have been discovered around the world. Where the tectonic plates pull apart, freeing the hot magma below to mix with seawater above, superheated reactions erupt from the seabed. Over time, they form gnarled chimneys composed of elements such as iron and manganese. The particular composition of these 'black smokers' depends on the rock that the seawater percolates through, picking up trace elements before it is superheated by the intense hydrothermal activity. Likewise, the ecosystems that they furnish vary around the world. Some are dominated by little-finger-long pale shrimp that nourish their bacterial symbionts inside their shells, making sure they are close enough to the superheated water to ensure an adequate supply of hydrogen sulphide. With a unique pair of light-sensitive 'organs' on their back, these eyeless shrimp can detect the infra-red radiation of 350°C water, a sensory system that helps them avoid getting too close. Each shrimp is torn between two powerful urges: the need to feed its microbial partners and its innate hesitancy to crawl towards a furnace.

In the vents found in waters surrounding Antarctica, a species of squat lobster farms microbes on its claws and on the underside of its carapace. Growing in thin filaments, they then clip their crop and feed themselves directly, a set menu strapped to their chest. While they are purely for nutrition, these microbial mats have the general resemblance of hairs, persuading the scientists who discovered this species in 2006 to coin the common name 'the Hoff crab'. I've never seen *Baywatch*, but I'm pretty sure the hairy chest of David Hasselhoff (aka 'the Hoff') has a very different purpose than that of this deep-sea crustacean. In a 2012 interview, Hasselhoff mentioned that he was sent a photo of the crab by his daughter, adding that he thought it was 'very funny, very cool and endearing'. I wonder if he knew that the crab lives in a world that smells like farts and that, from a distance, they look like a pile of skulls crawling over the seabed.

Shifting further north, into the Indian Ocean, and we find perhaps the most bizarre animals – yes, more bizarre than hairy-chested lobsters or giant, sulphur-breathing tube worms – that inhabit hydrothermal vents. 'It has this really punk rock look,' says Julia Sigwart, a mollusc specialist (or malacologist) based at the Senckenberg Research Institute, 'like an almost aggressive kind of animal, armed and ready for battle.' First discovered in 2006, the scaly-foot snail is adorned with what look like blackened sheets of metal, a skirt that wouldn't be amiss as a gladiator's finest garment. To drum up media attention on a humble mollusc, it has also been called a 'dramatic dragon-like animal' in scientific papers and dubbed the 'sea pangolin', a name that not only references the keratinous scales of this mammal group but also their endangered status. The shell's mineral coating is the colour of charcoal, more ammonite fossil than garden snail. Sigwart hadn't seen anything else like it. And then her PhD student at the time, Chong Chen, took a look inside these snails and its novelty only deepened. Like the giant tube worm, each snail had a massive trophosome, a palace for symbiotic bacteria. Every other part of their anatomy bent to this organ's whim. The powerful heart allocated most of the snail's blood to it. The large gills provided the gases it needed. And the brain was absent altogether. Only a very simple nervous system remained. 'The whole anatomy of the snail has actually been changed and contorted to fit what makes the bacteria happy,' says Sigwart. 'You can just imagine the bacteria sitting in there over a little steering wheel, driving the snail around, saying, "No, I want to be in this environment, closer to the vent, closer to the vent!"' she tells me in a nasal, high-pitched voice reminiscent of a Minion.

The scaly-foot snail is merely a vessel for bacterial desire. Seen through this lens, Sigwart knew that the skirt of armour had to be an adaptation of microbial, not mollusc, evolution. The idea of protection from predators was palatable to the media – 'it fitted with the punk rock narrative,' Sigwart says – but it didn't make sense in a hydrothermal ecosystem. 'There's almost no active predators in hydrothermal vent environments,' Sigwart says, 'because the environment is so hostile and toxic, especially if you're on the vent chimney very close to the hydrothermal fluids.' After leaving Sigwart's lab, Chen continued studying the scaly-foot snail with colleagues at the Japan Agency for Marine-Earth Science and Technology. There, he discovered that it wasn't armour plating, as expected. The scales were an outgrowth that acted like catalytic converters, soaking up the toxic by-products of a sulphur-based metabolism. Like with the turtle's shell, it was a form of armour, one adapted to deal with an inner enemy.

'I was like, *wow*, that's a completely different way of thinking about adaptation,' Sigwart says. 'And then you talk to microbiologists about this kind of system and they're like, "Yes, Julia, the whole world is controlled by bacteria. Did you not understand this?"'

To study the microbes within the animals that call hydrothermal vents home is a bit like describing a tree as a floating mass of leaves. What about the trunk? The roots? The connections that are formed between fungi and neighbouring trees? To understand the microbial population of a hydrothermal vent, scientists had to dig a little deeper into the subsurface. This is quite difficult when these ecosystems lie a kilometre or more below the ocean's surface.[2] The pressures are immense, the darkness complete. While drilling can be used to sample under the seabed, the advantage of working with underwater volcanoes is that they erupt. In what became known as 'snow blower' events, the microbes are emitted from recently exposed cracks in the seafloor, a cloud of microbial matter that can be surveyed to see what was living underground.

In 1998, such an event happened off the coast of Oregon, western United States. Known as Axial Seamount, it is an underwater caldera that has two rift zones – places in the seafloor that are being pulled apart – running north and south of its position, a loop of 'boundary faults' on three sides. In other words, this was a very active place, geologically speaking. The tectonic plates were both pulling and twisting at Axial Seamount. When a 30-centimetre-wide crack started to blow its microbial snow into the deep water, the Remotely Operated Platform for Ocean Sciences (ROPOS) was there to take a sample. The size of a small car and the weight of a Hummer H2, ROPOS used a vacuum-like tool to extract a litre of vent fluid, pulling it through a filter that caught the microbial matter in its finest mesh. Returned to the deck, the filters were frozen in liquid nitrogen for later analysis. Back at the laboratory at WHOI, Julie Huber compared the genetic sequences of her samples with the known types of bacteria and archaea to find what was living in the rock underneath Axial Seamount. There was a huge amount of diversity, 200 or more different types of bacteria. Some liked it hot, others cool. Some ate carbon dioxide and hydrogen and released methane and water, others metabolized sulphur.

For roughly 500 metres *below* the seabed, microbes were being swirled and mixed in an elemental meeting of fire, water and rock. The Azoic Hypothesis, once again, had been overturned: even the rocks underneath the ocean were a riot of life.

When I met Huber at her office at WHOI, she had been studying the ocean's subsurface for over 20 years and was about to set off on another research trip at sea. Outside her office, a large black container held all her sampling devices, rolls of sticky tape, labels and, on top, a big bag of pretzels. They are the best thing to combat seasickness, she tells me. After the pretzel hack, she told me that what is central to her work is understanding how life adapts to darkness. The discovery of hydrothermal vents, the descent of *Alvin*, inspired her to think about the places on Earth once thought to be too hostile for life. Hydrothermal vents, including the crack at Axial Seamount, were known as 'windows into the subsurface'. Huber was very keen to take a look through as many of these portholes into the unknown as possible.

This includes vents in our solar system. In 2006, three years after her first paper on Axial Seamount was published, the Cassini mission encircled the moon of Saturn – Enceladus – and photographed huge spouts of ice erupting from its southern pole. Receiving only a hundredth of the sunlight as our planet, this moon was seen as a frozen and lifeless sphere, a presumption of astronomy that led all the way back to William Herschel who first spotted this moon on 29 August 1789. The Cassini images told a different story: Enceladus was geologically active. Tugged and pulled by the gravitational forces of its lunar neighbours, the 20-kilometre-thick sheath of ice hid a liquid ocean that could be ten kilometres deep at the southern pole.

Later calculations estimated that the water that erupts from this moon's surface shoots out at the speed of a jet plane and reaches hundreds of kilometres into space. (While some falls back to Enceladus, this water also contributes to one of Saturn's famous rings of ice.) In 2008, NASA engineers sent new coded orders to the space probe, guiding it back around Enceladus and through the plume of these icy spouts. Just like ROPOS sampling a litre of subsurface fluid from Axial Seamount, Cassini had a taste of Enceladus's insides: a salty ocean. High pH, not too far off drain cleaner or bleach. Very little, if any, oxygen.

Julie Huber walks with me down the corridor in her department building, past the black crate with the bag of pretzels, and into the laboratory where her students and postdocs work on microbes from the subsurface here on Earth. It's empty at this time, just after 9am on a Monday, but their experiments and specimens are still scattered over their desks or stored away in incubators and fridges. Huber opens up a glass-fronted cabinet, and picks out a test tube that contains a couple of finger widths of pale liquid. Watered-down milk, I think. Actually, this is science's best guess at what Enceladus's oceans contain. As she holds the test tube, my head swirls with the notion that Huber is holding a piece of an alien planet, as casually as someone holds a glass of champagne.

This cloudy liquid is a concoction made by Sabrina Elkassas, a graduate student who shows me around her workspace after I've said goodbye to Huber. On her lab desk, among neatly stacked test tubes with colourful labels,

Elkassas shows me her A4 notebook that includes the recipe for Enceladus's ocean. She called it 42a, a reference to *The Hitchhiker's Guide to the Galaxy*, in which the number 42 represents all meaning: 'the Great Question of Life, the Universe and Everything'. The 'a' in 42a is also important. This was her first attempt, and it worked. Collaborating on the recipe with Tucker Ely, a scientist at Arizona State University who combined all the models of Enceladus's oceans into one best guess, Elkassas was expecting the mixture to simply form a mass of undiluted particles. Such 'precipitation' is common at these high pH values, a sign that the chemistry has compatibility issues. 'Everything just crashes out of solution,' Elkassas says. 'But this Enceladus one is the only one where I was able to get the highest pH out of any of the media… and [had] no precipitation. It's kind of insane that it worked that well. That never happens.'

What's truly insane, I think, is that she has introduced microbes into this solution and they are happily growing. The test tube that Huber showed me wasn't a lifeless liquid, it was a microbial world. These aren't extraterrestrials, however. They were taken from a deep fault line known as the Mariana Forearc, a place of vast plains known as mud volcanoes. Hidden under the surface of the seabed, these hydrothermal ecosystems don't erupt in vents but bubble at a much lower temperature. This particular mud volcano is five kilometres thick and stretches for 500 kilometres. While black smokers erupt a few metres from the seabed, mud volcanoes are spread over a huge area.

Elkassas shows me a few samples that she took during a recent research trip to the Mariana Forearc, a few cylinders of wet rock inside small plastic tubes. Most obvious is their colour: bright blue, like some sort of bedrock from Smurf city. Then she opens one of the screw lids and asks me to have a sniff. Dead fish. 'That's TMAO,' she says, referring to the organic compound we met in Chapter 5. Another: rotten eggs. 'Now you know what the ocean's subsurface smells like.' Knowing that the Mariana Forearc is the closest analogue to Enceladus's oceans found on Earth (high pH, low oxygen, lots of rocky minerals), I wonder whether this distant moon has a similar pong. It's probably best that the only visitors we can send are automated rovers and other vehicles that can penetrate through its ten-kilometre-thick icy crust.

Elkassas is yet to get a firm identity on the microbes she has been culturing, but their presence makes me ponder the words I use when talking about extraterrestrial life. Given the tenacity of microbes here on Earth, their ability to grow in some of the harshest environments, it's no longer a question of whether Enceladus is habitable. It's a question of whether it is *inhabited*.

Before I leave WHOI, Huber tells me that she thinks any extraterrestrials on Enceladus will be microbial, a world of single-cells. But I can't help but imagine a place of unknown animal patterns and body forms, a place where evolution had another blank canvas to express its inner strangeness in complete darkness. If there's one thing we have learned about hydrothermal ecosystems, it is that the violent collision of seawater with molten rock is a recipe for life. There are Gardens of Eden in the ocean's darkest places.

The subsurface under the oceans is dwarfed by the subsurface world of land, a place where rock begins under a thin layer of soil and extends all the way down to the magma that swirls around the Earth's core. In terms of the volume, this subsurface ecosystem is twice the size of all of the oceans combined.

While Huber was sampling some of the first deep-sea eruptions in the late 1990s, the terrestrial subsurface had a much longer history of scientific intrigue. There were two main ways to enter this underworld: a very large drill or following the tunnels made for mining. Most often, the search for life deep under our feet followed the pursuit of oil, rare minerals and gold.

The first discovery of life deep in the subsurface was blurred by this very unscientific method of sampling. Contamination was always a problem. Any microbe that was found deep in the rocky mantle could easily be a surface microbe that had either attached itself to the drillbit on its way into the rock, or it could have latched onto the specimen as it was brought to the surface and into the laboratory. In 1928, Chas Lipman, a Russian-born microbiologist working in California, claimed to have found bacteria 'in a [two-million-year-old] Pliocene rock which derives from a depth of several hundred feet from which it has recently been uncovered'. Incredibly deep and unbelievably ancient, this subsurface discovery was shadowed by Lipman's second sample: a lump of *650-million*-year-old rocks from the Pre-Cambrian, also containing microbes. Throughout the middle of the twentieth century, similar findings – often hailing from salt mines – would sprinkle the scientific literature. Ammonia-rich soils from Harlech, Wales; Silurian salts from New York State; Permian potash from 1,200 metres below Boulby, northeast England. All contained microbes and were aged between 650 and 200 million years old. '[M]icro-organisms,' one paper put it, 'given the right conditions, can exert a very tenacious grip on life.'

Studying these subsurface microbes was also like travelling back in time. One researcher working in the 1960s sat stunned as his Pre-Cambrian sample contained living *Bacillus circulans* cells. Never taking his eyes away from his microscope, Heinz Dombrowski wrote, 'it starts the first cell division after being dormant for more than 650 million years'. For a microbiologist, it was like collecting a fossilized dinosaur egg and then watching it hatch.[3]

Adding to the perennial contamination problem was the deafening level of disbelief. The extraordinary claims of ancient microbes from the deep, Max Kennedy, Sarah Reader and Lisa Swierczynski wrote in the journal *Microbiology* in 1994, 'stifled further research for many years'. There was a general sense that nothing could live in rock beneath our feet.

Even as the number of studies from the subsurface increased, there was an equivalent, Newtonian-like rise in scepticism and criticism. One side pushed, the other held firm. 'It was extremely difficult to convince people that it was anything other than contamination,' Barbara Sherwood Lollar, a geochemist who started her forays into the subsurface in the 1980s, recalls.

Sometime in 1996, Sherwood Lollar received a call from Tullis Onstott, one of the rising figures in subsurface microbiology. He had an idea of how to break through this impasse. 'Let's go to South Africa,' she remembers him saying. 'Let's start work there.' With active gold mines that descended four kilometres into the earth, they would enter a world that was heated by the geothermal activity below. At temperatures of 40°C or more, they would search for thermophiles, heat-loving microbes that couldn't live on the cooler surface above or in the lab. 'In these very high geothermal gradients', in Sherwood Lollar's words, perhaps their peers would be more willing to accept that these specialized microbes weren't from cross-contamination. 'It was an inspired choice,' Sherwood Lollar says, her words buoyed with the memory of Tullis C., or 'TC', Onstott, a colleague and friend, who died of cancer in 2021.

From these deep mines, they found microbes living in carbon-rich rocks heated to 60°C. Similar to those found in hot springs, this particular species of bacteria used iron to power their metabolism. Chipping fragments off the narrow shafts of the mine, Onstott turned on his Geiger counter and heard the constant crackle of background radiation. Packing the sample into a sealed, heat-proof bag and then a steel canister, he continued down into the mine. Only later would this moment form a major part of his theories on how life subsists in the subsurface. What if these microbes were using the power of radiation to survive in the darkness?

Returning to the South African mines in 2000, Onstott, Sherwood Lollar and their colleagues noticed that wherever they found radiation, they found hydrogen gas forming. 'I often refer to hydrogen as the jelly doughnut of the microbial world,' Sherwood Lollar says. 'If it's there, they're gonna eat it.' The gas they saw bubbling in the rock was formed when background radiation – from uranium, mainly, but also potassium – split molecules of water. Known as radiolysis, this wasn't anything new. A couple of scientists at Marie

Curie's laboratory in Paris formulated this basic equation in the early twentieth century. As well as hydrogen gas (H_2), other, more reactive, molecules can form. 'When you split water, you produce the hydrogen, but you also produce these highly reactive oxides,' says Sherwood Lollar. 'And it turns out when they attack the rock around them, they can liberate sulphate.' Like with the sulphur-slurping microbes on hydrothermal vents, these bacteria use sulphur to drive their metabolic machinery, just as we breathe oxygen.

It was a very simple equation, at least for biochemists. They call this form of life 'auto-litho-trophy'; essentially microbes that eat (trophic) rock (litho) and don't need any other input (auto). Not only were bacteria found deep in the Earth's crust, therefore, they were completely self-sufficient. They didn't need anything more than rock, water and radiation. 'You don't need light, food, or anything else from the surface. It contains everything it needs to exist, and it requires nothing from another organism,' Onstott said in an interview in 2012. 'Such things aren't supposed to exist. We thought all organisms depended on others, but this one doesn't.' We have rankings for the happiest places on Earth but perhaps this microbe is the most content organism alive right now. And if it won such an award, it wouldn't care one iota.

'Make no mistake, some organisms down there are what we call hetero-trophic,' Sherwood Lollar tells me. Some microbes in the subsurface are feeding on organic matter that either has come down in the water, or is present in the rock. In 2011, Onstott and his colleagues discovered a microscopic worm that was feeding on the bacteria that eke out a living, a species they named *mephisto* after Faust's Lord of the Underworld, Mephistopheles. During its 15 minutes of fame, *Halicephalobus* was called the 'worm from hell', given the deep, hot rocks it inhabited. In truth, this nematode worm has a great life. Warmth, all the microbes it could ever want. No predators or competition. This worm from hell is living in its version of heaven.

'As you go deeper,' Sherwood Lollar continues, 'you get into systems that are both ancient and [completely] isolated from the surface.' Some may have been isolated for a billion years or more. (As we talk, she says she could offer me a glass of water that's a billion years old, although she wouldn't recommend it.) Many of the early discoveries were indeed mired by contamination, but many, it turns out, were accurately revealing the extraordinary worlds that they claimed. In one obituary following Onstott's death, his former colleague Duane Moser wrote that such deep subsurface ecosystems were 'disconnected from the entirety of traditional biology, plodding along beneath our cities and politics, utterly indifferent to familiar perceptions of time; where seasons, ice ages, and even the continents themselves come and go without notice or consequence'. He called it a 'microbial *Jurassic Park*'. Sherwood Lollar, meanwhile, sees this 'deep biosphere' as a place comparable to the oceans: unfathomable, full of life for kilometres below the surface, largely uncharted.

And underappreciated. The microbes that live under the seabed may account for a third of all the biomass on Earth, a vast reserve of that life-making element: carbon. In a time when we are releasing carbon at unprecedented rates, the deep biosphere is a place where stores are kept out of our reach.

Hydro-geological separation: the independence of life from surface water and surface rock. While the deep biosphere is the largest shift in how we think about the Earth's habitable zones, it's not the only place with such independence. In 1986, a similar situation was discovered just 18 metres below some grassy knolls in Romania. Two kilometres off the Black Sea coast, with few tourists and residential homes, the land near Mangalia had been the potential location for a new power plant. While checking the bedrock for faults, one of the drills broke into a cave system, the noise of metal grinding through stone coming to an abrupt halt. With the risk of collapse, the factory would have to be built elsewhere. Before the drills moved out, however, Cristian Lascu, a speleologist working in Romania, investigated this newly discovered cave system. Unknown to him at the time, he was stepping into a world that hadn't had any input, never mind visitors, for over two million years.

The initial tunnel wasn't too different from any of the other caves he had studied. The oxygen level was quite normal. There was very little nutrient influx. It was very humid. Then the tunnel dipped into the groundwater. Returning with his scuba equipment, Lascu found that there were several air pockets that he could surface into, the undulations of the cave moving above and below the water line. It was these 'air spaces' that were truly remarkable. Even here, he couldn't breathe without the canister on his back. There was only 7 per cent oxygen, a third of that at the surface. It stunk of rotten eggs, the calling card of hydrogen sulphide. As with hydrothermal vents, it was a world shaped by sulphur-loving bacteria. Growing in mats on the walls and floating on the surface, they were the foundation of a rich and diverse group of animals.

There were no giant worms or hairy crustaceans. The inhabitants of 'Movile Cave', meaning 'hillock' in Romanian, are more familiar: leeches, snails, woodlice, spiders, scorpion-like creatures and centipedes. Familiar patterns and shapes of life, but special in their own way. As of the last survey in 2020, 37 of the 57 species that live here exist nowhere else on Earth; their distribution is entirely within the few cavities of hypoxic air and sulphur-rich water. At a length of just 21 metres, it's a pinprick in the Earth's biosphere, a miniaturized safari with a ten-centimetre-long centipede as the apex predator.

In their first study on the cave, published in *Science*, Lascu and his colleagues found that the chemical make-up of these animals supported the idea that they were dependent on the bacteria alone, not the chemistry of the surface world. While all life is based on carbon, there are different forms of this element – known as isotopes – that can weigh more or less, depending on where they come from. Carbon from plants, for example, is usually heavier in its atomic mass (known as C^{14}). Carbon from chemosynthesis – from the sulphur-loving bacteria, for example – is usually lighter (C^{13}). By measuring the ratios of these isotopes from the animals in Movile, Lascu and his colleagues were pretty sure they were not eating anything relating to plant matter. Their world was entirely based on the microbial mats.

It wasn't definitive, however. One line of evidence can be a fragile base to any theory. But there were other avenues that seemed to point in the same direction. For one, the geology above Movile Cave was layered with clay, an impermeable barrier. Second, the same year that the cave was discovered, the Chernobyl nuclear disaster spewed radioactive particles across Romania and other countries that neighbour Ukraine. The caesium and iodine from this invisible cloud was found everywhere, especially in water that moves across large expanses of land on its journey to the sea. They were never found in Movile Cave. Just a few metres below the grassy plains, this group of animals had found a refuge against one of the worst ecological disasters in history.

To protect this unique ecosystem, a concrete slab with an air-tight door was fixed to the place where the drill first spiralled into the ground. It is only opened once or twice a year to let scientists from around the world see, smell and sample this biologically indifferent ecosystem. The latest studies estimate that it has been isolated for over five million years, a time when our branch on the tree of life – *Homo* – had only just started to grow. Over the coming millennia, our species would harness fire, encircle the globe, settle into agriculture, discover the power of petroleum, and begin to warm and poison our surroundings with greenhouse gases and unstable elements. In darkness, the animals in Movile Cave slurped up their microbial crop and continued to diverge from their ancestors.

Let's return to the surface. Pull your imagination from the strata that lie deep in the Earth's mantle, the lines of rock laid down over millions of years of geological history.

If you're in the South African mine, its narrow tunnels lit only by your headlight, make your way back up, slowly. Re-emerging into the open air, sunlight tickles your retina once more and life seems richer after just a few hours in its absence. Plants are greener, birdsong sweeter. The breeze flows over your skin and feels ticklish, playful even. We are animals that rejoice in sunlight and find darkness threatening. We light fires to keep the night at bay, encircle campfires as our planet orbits the sun.

There are animals that live by the opposite rhythm. Sunlight, to them, is threatening as it exposes them to predators or stifling heat. While we are diurnal, they are nocturnal. To live in the absence of the sun is a place and time with riches and, often, fewer competitors. There's a bee in Panama that has adjusted its life cycle to the night-time because, in this biodiverse rainforest, there are so many other bee species busily foraging in the daytime. Around the world, millions of songbirds pass silently through the skies at night, a trick to avoid daytime predators while keeping cool as their muscles burn. In the tropical jungles of Indonesia, small primates called tarsiers use eyes larger than their brains to catch the few photons reflected from the moon's surface – two spheres in communion, one celestial, the other biological. (The science writer Matt Simon describes the tarsier's humongous, unblinking eyes as a 'general oh-criminy-did-I-leave-the-oven-on appearance'.) Such an adaptation for seeing in the dark is standard for night-time dwellers – just as big telescopes can see further into space, bigger eyes catch more light at night. If you can put a reflective layer of cells behind the retina, as in big cats like lions, then all the better. But, one summer evening, I was hoping to catch a glimpse of an even more extraordinary night-time resident, an animal that lives at the extreme end of the vision spectrum. Using only the few photons from distant stars, this creature isn't just able to see but see in *colour*.

It was nearly 10pm, and I was sitting in my garden with a tub of ice cream and a spoon. In order to cope with a two-year-old's early starts, I was usually in bed if not asleep by this time so this wasn't a regular activity of mine. I sat watching the honeysuckle flowers that crept over the stone wall from the neighbours' garden. In bright trumpet-shaped blooms of yellow and white petals, its smell was sweet, like vanilla mixed with jasmine, and thick in the air at this time. Behind the honeysuckle were a few pink roses the size of oranges. Bamboo shoots emerging like masts from the thick and unkempt hedge of ivy and bramble.

I couldn't see any of this in detail. The sun had set half an hour ago and my surroundings were lit only by starlight. I could see outlines of the wall, the leaves and the shape of the roses, but everything was monotone, a gradient of grey. I knew the colours and the flowers from memory, not sight.

While I wondered whether to eat the entire tub of ice cream and call it a night, a loud thrumming sound came over the wall, like a tiny wind-up aeroplane hitting its wings against each other. I sat up, a little in shock that only my second night of vigil might be fruitful. I put the ice cream down on the table and stood up for a closer look. It was definitely a moth. With a wingspan of a few inches, it was the largest moth I had ever seen, an insect the size of a small bird. But then my senses could glean no more; what species was it? I couldn't see any details other than two light stripes down its flank. I thought about catching this welcome visitor with one of my daughter's crab nets but didn't want to damage it. And so I let it thrum from flower to flower, and then back over the wall into the darkness beyond. My ice cream still uneaten, a green crab net in my hand, I stood stunned by an animal that is so common but so rarely seen.[4] The elephant hawkmoth lives in a dimly lit world that almost no other eye can penetrate.

In daylight, *Deilephila elpenor* is a flash of neon pink and lime green. While large for an insect, its common moniker doesn't come from its elephantine bulk. It is from the trunk-like shape of the moth's head and abdomen, thick at the top and tapering to a blunt point at the base. Imagine a tiny elephant trunk painted for a 1990s rave. (There are even larger hawkmoth species such as the privet hawkmoth with a wingspan of 12 centimetres, the same as a small songbird.)

The vibrant colours and size of hawkmoths make them a popular choice in butterfly houses or, more accurately, lepidopterariums. But the species chosen are usually active in the day (diurnal) so that they can be appreciated. Hummingbird hawkmoths are particularly popular. Elephant hawkmoths are active from dawn (crepuscular) into the darkest of nights (nocturnal). This is when honeysuckle, their favourite flower, perfumes the air with its strongest scent. Following the scent trail to my garden that night in early June, the hawkmoth would have seen the white and yellow petals of the honeysuckle, the pink of the roses. While the cells in my retina have sacrificed colour for sensitivity in the night-time, a common trait to all mammals, this insect was the first species known that can maintain colour vision in darkness.

Almut Kelber discovered this ability in 2002. As we spoke via video call, over 20 years later, I could hear the slight buzzing of wings in her living room in Switzerland. She has kept hawkmoths for decades. 'We love them,' she tells me, laughing. 'We just treat them as pets.' Previously at the Vision Group in Lund, Sweden, she extended her hobby of keeping hawkmoths into a groundbreaking scientific discovery. While colour vision in diurnal hawkmoths and butterflies had been proposed since the early twentieth century, Kelber was the first to prove that this was the case. Published in *Nature*, the first sentence of her paper was a comment on our long history of anthropocentric blindness: 'Humans are colour-blind at night, and it has been assumed that this is true of all animals.' But by giving elephant hawkmoths artificial flowers to feed from, Kelber found that they could decipher yellows and blues at light intensities that were as low as dim starlight.

To test an animal in the laboratory, aquarium or zoo, you first have to train it. And, even for the smallest of insect brains, this isn't difficult when there is an ecologically relevant reward. For hawkmoths, a sweet treat is the best offering. Inside her experimental flight cage, Kelber used a small capsule of sugary water placed behind pieces of yellow card – simple, simulated flowers. Extending their proboscis through a small hole in the card, the hawkmoths could slurp up the sugar water like the rich nectar inside their favourite honeysuckle blossom. Alongside these rewarding yellow 'flowers' were eight pieces of card in differing shades of grey, each lacking a sugary treat. In the relative bright light of dusk, a time of day when even humans can still see in colour (albeit badly), the hawkmoths quickly learned that yellow was their favourite. Given a choice in the future, they would choose the yellow card.

Then, by adding filters to the mercury bulb above the flight cage, Kelber dimmed their surroundings from dusk to starlight. The light levels were now 100 *million* times lower than a sunlit day. To her eyes, the yellow pieces of card lost their vibrancy and started to look the same as some of the grey pieces of card. In this monochromatic view of the world, yellow and one particular shade of grey were the same intensity – indecipherable. But the hawkmoths, all six of them, still fed from their favourite yellow flowers. The moths actually performed better at this level of darkness than under the brighter simulations of moonlight and dusk. With only starlight to guide them, elephant hawkmoth eyes have a decent claim to be the most 'out of this world' organs on Earth. To find their favourite colour of flower, they are guided by the rays that have travelled for at least *four years* through the vacuum of space.[5]

Rather than depending entirely on her own vision as a comparator, Kelber tested six adults working or studying at the university. None were colour-blind. And they weren't crammed into the flight cage or tempted with a sugary reward. Allowing their eyes to adjust, their vision was excellent at selecting the yellow discs at light intensities of early dusk, late dusk and moonlight. Even under starlight, the volunteers could choose the yellow discs correctly when compared to grey discs that were closer to black than white. But this wasn't colour vision. Their eyes were detecting the difference in light intensity, and yellow was seen as a lighter shade of grey that was easily recognizable. When only lighter shades of grey were placed next to yellow discs, the yellow flowers were only chosen 50 per cent of the time. Their choice frequency was, in other words, completely random.

To show that it wasn't just the lighter yellow flowers that hawkmoths could detect, Kelber repeated these experiments with blue discs, showing that they could still see these hues when paired with darker shades of grey. Again, humans failed to do this.

The results from these experiments were too good. By estimating the number of photons that hit the light-sensitive cells in the hawkmoths' retinas, Kelber and her colleagues concluded that the hawkmoths shouldn't have been able to do what they did. For all visual systems, whether the photosensitive cell of a tardigrade or the basketball-sized eyes of a giant squid, the detection of light is imperfect. The signal – photons of light that stimulate cells in the retina – is always tempered by 'noise'. This noise isn't sound but the amount of random firing of cells that happens in all biological systems. Think of it as the pops and clicks of a record player, the unwanted signals that come from imperfections in the grooves of the vinyl. In the retina, this misfiring of cells can send false reports to the brain. In daylight, they are hardly noticeable to our eyes. But when light levels drop towards darkness, they start to become more noticeable; we begin to mistrust what we see. In the record player analogy, turning down the volume brings the cracks and pops to the fore. In scientific parlance, this is known as the signal-to-noise ratio. It sets the limit to vision. Once noise becomes greater than signal, we can't navigate the world with any accuracy.

In the hawkmoths, however, this wasn't the case. The amount of photons that were hitting their photoreceptors was lower than the estimated background noise. Misfires outnumbered the meaningful. 'Reliable discrimination of blue from grey is impossible under these conditions, and this is especially true for the training blue and the medium shade of grey,' wrote Kelber and her colleagues at Lund University, Anna Balkenius and Eric Warrant.

Do hawkmoths break the biological rules of vision? Are they guided by some quantum trickery that we can't explain? Not quite. Part of the answer, it turned out, was that their retinas are able to integrate any signal for a longer time, seeing more slowly until they have detected enough photons to make sure it is real. Known as summation, it is very similar to a camera set to a low shutter speed, a technique that can take in enough light to create an image. 'The visual system adapts to using different, so to say, pixel sizes,' Carola Yovanovich, a vision researcher who studies the incredibly sensitive eyes of frogs, tells me. 'When it's very bright, you can afford to have very small pixels and a very high-definition image because you have enough light for each pixel. And when there's less light, if you keep such a small pixel type, there are many pixels for which there will be no light at all, because there are a few photons falling there. Many pixels will end up empty. So you can say, OK, I will combine these four pixels into one such that at least some light gets there. And, of course, you will lose resolution, sharpness in the image, but at least you'll see something.' Summation like this is how photographs of the Milky Way are taken, how neon green trails of fireflies are captured in North American forests, a means of catching as much light as possible from a dim source. The eyes of hawkmoths have hundreds of lenses all directing light to each photoreceptor, thereby amplifying the signal to overcome the noise; to see colourful flowers in the dark.

'I don't think it's the whole picture,' says Eric Warrant. 'I think we've come part of the way with summation. I just don't feel it's enough still; we haven't solved this mystery entirely.' While he supervised Kelber at the Vision Group in Lund, Warrant's main interest over the years has been nocturnal bees. They navigate from nest to flower under the darkness of a jungle canopy, guided by a mere five photons hitting their photoreceptor cells every second. 'That's nothing,' Warrant tells me. What's more these bees still have the eyes of a daytime bee. They don't have the many-lenses-to-one-photoreceptor of hawkmoths, just the standard compound eyes of day-active insects (such as houseflies) in which every photoreceptor has just one light-funnelling lens. 'Their eyes are hundreds of times less sensitive to light than the eye of a hawkmoth,' Warrant says. 'But these bees have been forced into this nocturnal lifestyle in the rainforests because of competition, predators and parasites. So they've taken advantage of nocturnally flowering plants more and more, despite having this incredibly unsuitable eye.

'Not only are they forced to do it,' Warrant adds, 'but they do it and they get away with it, and it works.' To survive and thrive in a busy world, animals can adapt to places that, at first glance, can seem completely unsuitable.

Since the discovery of hydrothermal vents in 1977, life in darkness hasn't just emerged as a minority corner of the biosphere but has exploded in technicolour.

CHAPTER EIGHT

TASTE OF A POISON PARADISE

Radiation

With the gentle buzz of honeybees at our feet, we carefully step between clumps of purple heather while keeping our eyes on the four powerful horses in front of us. All young males that only stop grazing to shake their heads or nibble at their flanks, they are Przewalski's horses – one of the oldest and most feral of horses[1] – living in Paleolítico Vivo, a prehistoric nature park in northern Spain, a place where human history runs deep and the current owners of the land are trying to recreate a snapshot of the Pleistocene. There are European bison with bulbous shoulders and scruffy manes, cows with large horns that are being bred to look and behave like aurochs, and two types of horses: the Przewalski's that I'm watching and the grey and brown tarpans that look like a lean and shaggy version of any racing breed. The colour of freshly cut straw with thick brown manes, Przewalski's are reminiscent of ponies with oversized heads, their powerful jaw muscles pulsing with every trimmed patch of grass. Germán Orizaola, a softly spoken ecologist from the University of Oviedo in northwestern Spain, is here to collect their poo.

When I agreed to visit him, driving two hours from the airport in Asturias the day before, I thought he needed just one sample: a horse defecates, we pick it up, we leave. But I soon learn that he has to collect samples from all four horses. When I ask how long it usually takes, Orizaola, his voice barely audible above the bees and the reverberating song of a skylark, gives me a range. Sometimes it's an hour, he says. One day he started at 10am and didn't finish until three in the afternoon. With a potentially long day ahead, I put on some sunscreen and ration my small bottle of water. It's nearly 11am and the sun is beginning to break through a light cover of clouds. The only sound other than the crunch of our footsteps, the bees and the skylark is the very loud HEEEE-HAWWWW of Romulito, a very old donkey who thinks he is a horse.

We follow the small herd as they graze through their open paddock, bordered by a dense stand of conifers at one side, a valley filled with gnarled oak trees at the other. Except for Romulito, the place is a simulacrum of a lost world. To Orizaola, however, it is a very handy stand-in for his study of Chernobyl.

Orizaola is interested in how life can adapt and thrive in the presence of radiation. Reclaimed by nature, the area surrounding the nuclear power plant near Chernobyl in Ukraine is home to wolves, wild boar, deer and, since 1998, herds – or, more specifically, harems – of Przewalski's horses. Buried in the soil under their paws and hooves, the radioactive particles that were emitted after the explosion of reactor 4 on 26 April 1986 can still make a dosimeter crackle like white noise, the sound of charged particles firing into the surroundings. If they penetrate into the body of an animal, they can shred DNA in several ways: the double helix can be severed, its two ladder-like struts broken in what's known as a double-strand break; a single rung of the double helix – a base – can be blasted from its proper location, a 'single-point mutation'; and then there are lesions, swellings around a particular point of the DNA that was damaged by the emissions of a radioactive particle.

And yet, even with all these textbook examples of radiation injury, life around Chernobyl is thriving. From their initial reintroduction population of 31 individuals, the latest survey that deployed camera traps throughout the exclusion zone – together, capturing 411,000 photos – estimates a healthy population of 150 individuals, a five-fold increase in two decades. Between 2016 and 2019, Orizaola walked around the southern part of Chernobyl's exclusion zone for two weeks every year, without a protective suit or PPE of any kind. (The southern part of the exclusion zone is less radioactive since the prevailing winds in the spring of 1986 were northwesterly, blowing much of the radioactive debris towards Russia and Scandinavia.) His dosimeter crackled now and again but the dominant sound was that of nature: birdsong, frogs croaking, the rustling leaves of three-metre-tall oak, aspen, birch – trees that germinated in the soil as the power plant exploded in a flash of blue light and fire. As he noted in his blog in 2019, 'wild boar, roe deer, a red squirrel, hares, and at one place dung with the look of a big cow, maybe from a European bison? Among the birds, we saw black stork, black grouse, honey buzzard, swans, black terns, golden orioles, cuckoos, reed warblers… and more.'

'It's an extremely beautiful place,' he tells me over our initial video call. The most dangerous radionuclides (radioactive particles) – iodine, in particular – have long disappeared, and the exclusion zone is now a place of longer-lasting but weaker ones such as plutonium that, Orizaola tells me, you can hold on your hand without harm. Only when ingested can the radiation ricochet into your internal organs. When outside the body, the skin is too thick a barrier.

'There is less than 10 per cent of the radiation that was released at the accident that remains in the area,' he tells me. 'Ninety per cent of it has disappeared.' To put these numbers into context, he adds, his two weeks in the Chernobyl exclusion zone expose him to the same dose of radiation as a return trans-Atlantic flight from Madrid to New York.

Scientific interest in radiation is often biased towards the extreme levels emitted from nuclear warheads or exploding power plants. We know a lot more about how acute radiation affects biology – badly – than we do about low-level, often chronic exposure. The former is intriguing for its morbid nature – tissues melting, flesh burning, DNA breaking – while the latter is rife with extraneous factors, confusing to pin down causation with any certainty. If someone dies of cancer at the age of 65, who's to say that being near Chernobyl in 1986 caused it? Orizaola is just one ecologist trying to shift our focus towards the longer-lasting and more environmentally relevant forms of radiation. Leave the explosive stuff for HBO, he might say, science is interested in the important story.

Przewalski's horses, Orizaola thinks, are the perfect animal to reveal how living in an exclusion zone affects the health of a large mammal. Only introduced into the area in 1998, the horses didn't experience the extreme levels of radiation that swept through the area in the late 1980s and early 1990s. While other research groups study feral dogs that were left behind after the evacuation of Chernobyl and the nearby town of Pripyat, Orizaola thinks that the horses are more informative, largely because they weren't in Chernobyl during the explosion. How does a large animal survive in chronic, low levels of radiation? His presumption: very well.

By collecting the Przewalski's poo, what Orizaola calls 'brown gold', he can learn a lot about the horse itself. Its parasite load can be counted. Parts of its genome can be sequenced. And, from these samples of DNA, Orizaola hopes to understand how these horses can deal with radiation that is still at harmful levels 40 years after the reactor exploded. Is there a common genetic toolkit that allows the damaging effects of radiation exposure to be reduced? Or, in a theory that he admits is hugely speculative, is the radiation actually helping these populations of horses to emerge from their near extinction a few decades ago? The entire population of Przewalski's can be traced back to 13 individuals returned to the wild from captivity in 1992, an ancestral pool with very low genetic diversity. With its mutagenic effects, is the radiation around Chernobyl providing them with a boost of diversity, making a vulnerable population more stable?

Whatever the case, whether we're speaking over video call, in the field or lab in northern Spain, or at a conference in Edinburgh, it's obvious that Orizaola sees radiation in a way most people don't – far from harmful, it is something that life has grown up with, an environmental variable like temperature or water. 'We used to think of radiation as something artificial and weird, and that animals are not naturally exposed to,' Orizaola says. 'But that's not the case. We are always exposed to radiation, there is radiation everywhere.' Ultraviolet rays from the sun, infra-red radiation from water heated to 400 °C at hydrothermal vents, the molecular decay of uranium or radon within bedrock, even the food we eat such as radioactive potassium in bananas: life has never been far from the power and potential harm of radiation.[2]

While radiation includes visible light – the blues, greens and reds that our retinas detect and mix – it is the emissions from radioactive atoms that we are most often drawn to. A product of the decay of unstable radionuclides such as radon, plutonium and uranium, it is called ionizing radiation because it has the power to split stable atoms into ions, stripping them of an electron and making them unstable and reactive. Found naturally in rocks around the world, these radioactive elements have been mined, purified and placed in an unnaturally cosy space inside nuclear reactor cores. At Chernobyl reactor 4, for example, over 20 tonnes of uranium-235 (or 235-U) was packed together and compressed into fuel rods. The number 235 relates to the molecular weight of the atom. Similarly, oxygen can be written as 16-O, sulphur as 32-S, and, if you're feeling very brave, roentgenium as 222-Rt. The only difference being that this list is of stable atoms, each with a balanced number of protons and neutrons inside its central core. Uranium-235, however, is not stable. The *element* of uranium is 238-U. The radionuclide, therefore, has lost weight. It has shed some of its core. It is decaying. The Nobel Laureate Marie Curie called this radioactivity.

As each uranium-235 decays, it sends out parts of its nucleus into the surroundings. Some of these emissions are larger than others and can cause more damage, a cannonball compared to a bullet. Alpha particles, beta particles, gamma rays: the terms aren't important, just that they provide a range of projectiles that can penetrate the radionuclide's surroundings. Some last longer and are weaker, halted by a leaf's cuticle or a layer of skin. Others are heavier and can force themselves through branch and bone. By placing these radionuclides together (say, in a series of closely packed uranium rods), this atomic artillery fire can cause neighbouring atoms to rupture. Nuclear fission, as it is known, releases the energy that boils water into steam that is then forced through a series of turbines. During the explosion of 26

April 1986, the fission reaction reached supercriticality and the pressurized steam exploded, sending the shrapnel of the reactor core flying into the surroundings. The power of radiation was unleashed on an unprecedented scale. And yet even this wasn't enough to create an inhospitable wasteland. Just a few years after the biggest nuclear accident in Earth's history, life not only recovered but rebounded.

Before we understood the potential power of ionizing radiation, we probed its secrets with little care for the consequences. Discovered in the late nineteenth century, X-ray radiation was known to burn the skin and so practitioners used this as a measure of the dose rather than an indication to stop. Radiographers would time the onset of erythema – redness, inflammation, burning. Then, a few years later, radium was discovered to glow with an intense green hue, and so watchmakers painted the substance onto the numbers and hands of clock faces. More specifically, the young women painting the dials used paint brushes that they licked to a point, each time adding a dose of radiation straight into their mouths. Ironically, their time was being cut short. Many of these women – known as 'Radium Girls' – would die from cancer, but not before their teeth fell out and their jaws disintegrated. Marie Curie, someone who knew more about radiation than anyone else alive at that time, noticed that the minerals pitchblende and torbernite were far more radioactive than the uranium that they contained. She theorized that there must be another, even more radioactive particle inside. Pursuing this unknown element in a run-down shed with no protective gear, Curie was breaking new ground in science while simultaneously breaking her own body. In 1898, she published her discovery of two new elements: polonium (named after her country of birth, Poland) and radium. Her death 30 years later came after years of weakness and, finally, aplastic anaemia, an inability to produce new blood cells in bone marrow, a common consequence of radiation poisoning. Many of the Radium Girls had died from the same disease.

Even in the middle of the twentieth century, a time when the harmful effects of radiation were being utilized specifically to design lethal bombs, there was still very little regard for personal safety. At 15:20 on 21 May 1946, Louis Slotin was working on a small reactor core at Los Alamos in New Mexico when the screwdriver he was using to hold the beryllium shield above the plutonium core slipped. As the two parts of a nuclear bomb met, the seven people in the room saw a blue flash emanate from the core as a wave of hot air washed over them, like an oven being opened and closed. Still holding onto the beryllium shield, Slotkin quickly ended the reaction by lifting it back up and throwing it on the floor. But this split second was already too long. After leaving the building, Slotin vomited – a common side-effect of radiation poisoning – and was rushed to hospital. Suffering from diarrhoea, blisters on his hands and arms, and radiation both inside and outside his body (a condition that one doctor called 'three-dimensional sunburn'), Slotin then experienced a 'total disintegration of bodily functions' and died. It was 11am on 30 May, just short of nine days after the accident.

The explosion at Chernobyl Nuclear Power Plant was similarly a result of human error (and a design flaw of this particular type of reactor), but the amount of radiation – 1,000,000,000 Curies – released hadn't been experienced on Earth ever before. The workers at the plant and the firefighters who arrived a few minutes afterwards to try to douse the flames were being slowly shredded by an invisible force that could only be felt as a metallic taste on the tongue and a burning of flesh. Again, there was an underlying belief that life would endure. As Adam Higginbotham writes in *Midnight in Chernobyl*, some believed that 'Those men who allowed themselves to fear radiation were most at risk. But those who came to love and appreciate its spectral presence, to understand its caprices, could endure even the most intense gamma bombardment and emerge as healthy as before.' Some of the workers at the plant were told that radiation was so harmless that they 'could spread it on bread'. Over the coming weeks, their bodies would start to blister, burn and eventually melt. Fifty people died from this 'acute radiation syndrome'. An estimated 4,000 more would die, over the coming years, from various forms of cancer.

Radiation has a power that cannot be measured in Roentgen or Curies. Perhaps because of its invisibility, it can maintain a semblance of safety even when it is screaming through a person's vital biochemistry. Like magic or religion, it seduces and shifts depending on the stories that we tell. From Marie Curie's work on polonium to the Chernobyl disaster, people will put themselves face-to-face with this most powerful of forces. Whether awestruck or ignorant, unrepentant radioactive particles tear through a strand of DNA in much the same way.

Breaking through a piece of DNA – the double helix of life – sounds inimical to health. But it is a normal part of a cell's function and upkeep. Like a car in need of a service, DNA can be repaired. Detecting the change in genetic sequence, a cell's ever-present groundskeeping proteins either kill the cell, a sacrifice for the greater good, or reinsert the correct code. It is this process of DNA repair that allows some animals to survive even the most extreme levels of radiation.

One of the most remarkable examples is the humble *Hypsibius* tardigrade I found in a clump of moss near my home in Devon. In a paper published in 2023, Bob Goldstein, the tardigrade researcher we met in Chapter 1, used a shoebox-sized machine that could deliver a precise amount of radiation to samples small enough to fit in – microbes and tardigrades, essentially. 'It's this tiny thing that a human couldn't get into, but if you could you'd be dead in about two or three minutes given the dose of ionizing radiation,' he says. 'Tardigrades are in there for a day, two days. It takes two days to reach the stage when you kill about half of them.' Leaving the tardigrades inside this 'irradiator' for a day, Goldstein and his colleagues found that these animals do accrue a lot of DNA damage. 'You check the same animals again 24 hours later,' he adds, 'and they're fine. They've lost that DNA damage.' This response is found across the animal kingdom. The same – or similar – proteins detect the damage, dock onto the particular site of DNA, and get to work fixing the changed sequence. 'It's not like they've invented new wrenches and screwdrivers,' Goldstein says. 'They're using the same wrenches and screwdrivers, they just make a million of them.'

What these tardigrades are doing is producing these proteins to, in Goldstein's words, 'an astonishingly high level'. Compared to before their blast of radiation, they have increased by 300-fold in some cases. This DNA repair response is so overwhelming for *Hypsibius* that other basic cellular functions are paused. The production of many of the other crucial proteins that a cell might need is forestalled. 'To me, it's like in wartime [and] what we do with factories,' says Goldstein. 'A factory that was making shoes suddenly has to make munitions. It's a pretty dramatic change, way more than we expected.'

While DNA repair can be seen as a difference in degree, there is another way to survive extreme levels of radiation, one that prevents the damage from occurring in the first place. Tardigrades belonging to the *Ramazzottius* genus depend upon a gene known as Dsup – or damage suppressor – that produces a protein that essentially sticks to DNA and prevents it from breaking apart. To understand how, we need to know a little bit more about what DNA looks like when inside the cell's nucleus. The famous discovery of the double helix was the most basic shape that can define DNA. But this is not how this molecule is packaged in nature. 'It's not just naked DNA bobbing around,' Grisel Cruz-Becerra, a biochemist at University of California, San Diego, who has studied Dsup in tardigrades, tells me. 'Most of it is compact DNA, it's wrapped around histones, it's not quite so simple [as a helix].' This mass of DNA – all wrapped up and contorted inside its nucleus – is known as the nucleosome. And what Dsup seems to do, Cruz-Becerra says, is to bind to a specific site of the nucleosome and transform its shape from a disordered blob into one that protects the DNA from breaking. In a paper published in 2019, Cruz-Becerra, James Kadonaga and their colleagues at San Diego found that Dsup shares a small sequence of DNA with vertebrate 'HGMN' proteins, a small snippet of tardigrade resilience within us all.

What this structure is, is still unknown. What's clear is that Dsup can be used to protect biological material from damage. It has been inserted into plants, flies and even human cells in culture, and each time it extends the lifespan of these cells when exposed to stressors such as radiation or reactive oxygen species (those by-products of the metabolism that we call ROS). While the research into Dsup is still in its infancy, Cruz-Becerra sees its potential for biotechnology, specifically in forms of gene therapy that deliver new genetic sequences into the body to correct for harmful mutations. Currently, this form of medicinal DNA repair, however, is limited by the length of time the new gene can survive in our body. As with any virus or bacteria, anything novel is quickly destroyed by our immune system. With Dsup, perhaps this response can be buffered, and a beneficial DNA sequence has time to reach its destination.

Why do tardigrades withstand radiation so well? They would never experience the unfiltered rays of the sun, the irradiation machine in Goldstein's lab, or the emissions from an exploded nuclear power plant. It all links back to our first chapter on extreme adaptations: water stress. 'If you're adapted to survive desiccation, you probably have really good DNA protection and/or DNA repair,' Goldstein says. 'That's our best explanation.'

Repair and protect: these are the two principal means of surviving intense bursts of radiation. If we stray to another branch of the tree of life, however, a third option begins to reveal itself. It is subtle, still largely unproven, but the potential for a dramatic shift in our understanding of life on Earth is certainly there. Those ever-inventive fungi that feed on the decomposing matter of plants and animals might harness radiation as if it were an all-you-can-eat buffet.

In May 1997, over a decade after the Chernobyl Nuclear Power Plant exploded, Nelli Zhdanova travelled to the site to survey its surroundings. The trees where radiation had burst through like some galactic ray gun had scorched into a sea of bright red, as if the leaves were now burned for eternity. The reactor itself had been covered by a concrete shield nicknamed the sarcophagus. But when Zhdanova and her colleagues looked at the walls, ceilings and cable passages (places that weren't subject to chemical decontamination procedures) inside this shield, it was clear that there was life growing in this 'death zone'. In a room where only an eight-wheeled rover could enter, the black shading on one of the walls suggested some type of fungi was growing in a place where humans could still not enter. In the adjoining rooms, Zhdanova and one of her colleagues wore plastic hazmat suits that encased their bodies in a protective cocoon. Plastic shields covered their faces and a length of plastic tubing connected them to their oxygen tanks. Connecting to their rear ends it looked like they were aliens that breathed through their tails. Collecting her samples by pressing cotton buds on the walls and ceilings, Zhdanova placed each swab into a Petri dish filled with nutrient-rich jelly. Over the next two months, she let whatever she had collected grow. In total, there were 37 species of fungi growing inside the abandoned reactor. Most common was *Cladosporium sphaerospermum*, a species found in almost every sample. *Penicillium hirsutum* was found in nine out of ten places that were sampled.

'According to our calculations, the radiation doses received by these strains must reach hundreds of Sievert, as a minimum,' Zhdanova and her colleagues wrote. 'For comparison, between 2 and 10 Sv in a short-term dose would cause severe radiation sickness in humans with increasing likelihood that this would be fatal.' The international limit for nuclear industry employees is around 0.05 Sievert per *year*. These fungi would break this threshold in a few hours.

The implications of this discovery were largely unknown. 'Chernobyl is, fortunately, unique,' the microbiologists wrote. There is no comparable system, no level of radiation that matches the scale of this disaster. Instead, Chernobyl represented the extreme end of a spectrum of catastrophe. From Bikini Atoll nuclear sites to wastewater sampled from other nuclear power plants, there was a long history of finding fungi in the most radioactive places. Chernobyl, *even* Chernobyl, was habitable for these hardy organisms. Near the end of their six-page paper published in the journal *Mycological Research*, Zhdanova noted that 80 per cent of the fungi sampled were darkly pigmented, not dissimilar from the black mould that grows in damp corners of houses. With this observation, she made a prediction. 'There may be a correlation [of pigmentation] with their ability to tolerate such high levels of radiation, not only to survive but to grow actively for long periods of time.'

Back in her lab at the National Academy of Sciences of Ukraine, Zhdanova started to unpick one of the patterns she had seen in the field. The fungi that were found growing on the remnants of the graphite rods of the reactor seemed to have more hyphae – the very thin, root-like structures of fungi – growing next to the 'hot particle' than away from it. Perhaps the fungal species hadn't just landed on these structures as spores. Like plant roots extending towards water, they had grown towards them. In response to such notions of 'radiotropism', some scientists – including Zhdanova – wondered whether they were simply growing towards the graphite core because it contained carbon, the element that comprises the skeleton of a glucose molecule. And so Zhdanova removed this from the equation. Placing a few species of fungal samples from Chernobyl near a source of radiation that was physically separated by the lid of the Petri dish, she still found that they grew towards ionizing radiation in 86 per cent of cases.

Radiotropism, the directed growth of fungi towards ionizing radiation, was real. In 2007, Ekaterina Dadachova and Arturo Casadevall, a nuclear chemist and a molecular biologist respectively, added another layer to the story. The reason that these black fungi were moving towards something we consider to be lethal was because they were actively seeking it out. They weren't, in other words, just there because there was a lack of competitors or predators (amoeba eat fungi, for example). They came up with the term 'radiosynthesis', an extreme version of photosynthesis that is working with levels of radiation around a million times more powerful than sunlight. To harness such an explosive force, you have to reduce it into lower, more palatable doses. For this, they use melanin, the pigment that makes our skin brown in the sun and makes the fungi growing on reactor 4 black.

'The energy of ionizing radiation is really, really high. It's much more than any living organism, including a fungus, needs,' Dadachova says. 'So melanin is sort of a transducer of radiation.' Like reducing a high voltage plug into one suitable for Christmas tree lights or shavers, melanin provides the fungus's cells with a level of radiation that doesn't cause all its machinery to burn out. During his visits to Chernobyl, Orizaola, along with his student and colleague Pablo Burraco, also found that tree frogs living in areas with historically high levels of radiation were darker than those from outside the exclusion zone. While the pattern is only correlative, the fact that both fungi and frogs produce more melanin in areas where radiation has been highest is a strong argument that it has some important protective value to these organisms.

For fungi, however, melanin is both a transducer and a power factory, protector and provider. In their experiments with common fungal species that are studied primarily as pathogens, Dadachova and Casadevall found that the growth of fungi exposed to ionizing radiation was two and a half times higher than those that were not. Excluding all other nutrients and sources of energy, they made the bold conclusion that they had discovered a new means of energy acquisition in life. 'There are recent reports that certain life forms can utilize non-conventional forms of energy,' Dadachova and Casadevall wrote, referring to microbes in hydrothermal vents and in the hydrogen-infused deep subsurface. 'On the basis of these precedents, we cautiously suggest that [these fungi have] the ability to harness radiation for metabolic energy.' Photosynthesis, chemosynthesis, radiosynthesis – the three pillars of life on Earth, potentially.

Dadachova is now partnering with plant biologists to try to disentangle the intricate mechanics of melanin and how it may be on par with photosynthesis. But this is an enormous task for one lab and funding hasn't been forthcoming. The biggest barrier that Dadachova and Casadevall come across is disbelief. Discovering a third means of autotrophy, of life sustaining itself directly from an energy source, they thought, would be a one-way path to papers in *Science* or *Nature*. Instead it was published in a brand new open-access journal that was looking for slightly more radical research to make a splash. 'When you show people [life] growing on radiation, after having an entire life being told that radiation is bad for you, you just can't accept it,' Casadevall says. 'I remember one reviewer saying, "I can't find anything wrong with this, but you're at the level of religion. Believe it or not."' Again, radiation has a force that science cannot explain.

The most extraordinary feat of these fungi might actually be how very ordinary they are. 'They are everywhere,' Dadachova says. 'All those fungi [in Chernobyl], they come from the soil. And because many of them are capable of producing melanin, they started to take advantage of that environment.' They can also take advantage of our own bodies, infecting people who have weakened immune systems. Emerging from soil whenever the chance arises, they are called opportunistic pathogens. 'Fungal diseases are rather new diseases in medicine,' Casadevall says. Viruses and bacteria go back to prehistory but fungi didn't really become a problem for humans until the late twentieth century. And what happened in the late twentieth century is you have medical progress, made at the cost of immunity.

'You get transplant patients, you get cancer patients, and you have the HIV pandemic,' he says. 'So what you see as you weaken immunity is that fungal diseases then become a major problem.' The flexibility of fungi is one of evolution's finest achievements. But while impressive, we shouldn't forget the incredible suffering that these organisms can bring to people already suffering with disease.

Since publishing the hypothesis of radiosynthesis, Dadachova has also been working on the medical applications of melanin. In 2012, she found that by feeding mice a variety of black mushroom that wouldn't be amiss on a fancy restaurant menu, they were able to survive 'lethal doses of radiation'. Such a protective mechanism, if found to occur in people, could be used for anyone who is working near a radioactive source, for cancer patients exposed to high levels of radiotherapy, or for astronauts journeying into space. Then, in 2020, her lab found that using soluble melanin – i.e., no actual fungi needed – could similarly protect lab mice from harmful ionizing radiation long after the exposure occurred. 'We showed that even up to 48 hours after irradiation, it can protect the mouth and gastrointestinal tract from the effects of ionizing radiation and basically save those mice from dying,' Dadachova says. 'This is very important from the angle of, say, a possible mass exposure – God forbid – as a result of a nuclear accident or some kind of war. So you will have a huge number of people who need to take something because they've already been exposed. And there is not enough time to give it to them within the first couple of hours. So you have to develop some kind of mitigator which can help them, say, 24 to 48 hours after exposure.' Just as people in neighbouring countries to

Ukraine were advised to take potassium iodine to prevent radioactive iodine from being absorbed by their thyroid gland, future nuclear accidents might necessitate the consumption of certain strains of fungi.

Radiation on Earth is just one application for melanin-rich fungi. Discussing their work with NASA engineers, Casadevall and Dadachova have plugged the idea of using melanized fungi to protect astronauts in space. Rather than wrapping a spacecraft in lead or some other thick metal, a population of fungal spores could inoculate a cavity within the ship's exterior. Surviving off the waste from the shipmates, they could grow in a cocoon-like chamber on the outside of the vessel, a biological shield against the ravages of radiation. 'If we're gonna go to Mars,' Casadevall says, 'we've got to solve the radiation problem… That's the biggest impediment for humans [living] in space for a very long time.' For a group of organisms that are often seen primarily as infectious agents or potentially lethal foodstuffs, fungi might protect us over an arduous journey to an inhospitable planet. To reside on a planet without any atmosphere or protective ozone, humanity might be wrapped in a cocoon of hyphae.

To Germán Orizaola it's no surprise that life is thriving around Chernobyl.[3] The threat from radiation is far lower than the very recent threat – and disturbance – from humans. While we have been around for a few hundred thousand years, radiation has been around since there was a sun and an Earth. Life has been shaped by it, just as it has been shaped by water and rock.

Needless to say, our understanding of radiation has recently undergone a reversal. For much of the twentieth century, radiation of any dose – low or high, acute or chronic – was considered to be detrimental to health, increasing the chances of cancer and death. Even background radiation from radon was linked to increases in mortality. The fears over nuclear fallout from the Second World War, the Cold War and then the Chernobyl explosion added to the general sense that radiation was a bad thing. But since the 1980s, a growing number of scientists have provided evidence that suggests a slightly different story. Studies into single-celled organisms such as ciliates, algae or photosynthetic bacteria found that growth rates were higher when background radiation was supplemented with a low dose of ionizing radiation. One research group grew single-celled organisms – *Paramecium* and algae – at the Centre National de la Recherche Scientifique, a facility that included a laboratory underneath the Pyrenees mountains. Exposure to a small amount of radioactive cobalt (60-Co) increased their rate of reproduction while, crucially, shielding them from background radiation actually *reduced* growth. This study – one that was replicated by scientists in the US – begged the question of whether life requires a low amount of radiation to thrive. Rather than any and all radiation being bad for biological systems, perhaps there's an optimum dose – an intimate connection with the radiation-infused Earth that life evolved in.[4]

What's true for microbes has also been demonstrated in mice. In 2006, a study led by Kazuo Sakai at the Low Dose Research Center, Tokyo, found that laboratory mice that were irradiated with 0.5 Gray (1.2 milli Gray for 23 days) increased their production of antioxidants and were better at repairing their DNA compared to mice that weren't exposed to any radiation. A low dose of radiation, in other words, seemed to activate parts of the cell involved in protection and repair. Similarly, other studies have found that radiation can enhance immune system function, reduce inflammation and promote the death of cells that are damaged. When the mice in Sakai's study were then exposed to a larger dose of radiation, they showed significantly less DNA damage than their non-irradiated counterparts. Ionizing radiation could be less of a poison and more like a vaccine.

This dose-dependent effect of radiation has been termed the 'hormesis hypothesis', a more general idea that many toxic stimuli can have beneficial effects at low doses. But while there is laboratory evidence to support the hypothesis, there are also many studies that need to be taken into account. In epidemiological studies that track health over many years in large numbers of people, for example, low doses of radiation correlate with increased likelihood of blood cancer and heart disease, especially if the radiation occurred in utero or childhood.

When speaking of the harms of radiation (whether from sunlight or radioactive particles), humans are one of the exceptions. We live a long time. We breed late in maturity. We have active lifestyles long after having children. Women can expect to live 30–40 years more after the menopause.

Still, the hormesis hypothesis adds a tempting – if still unproven – epilogue to the story of Chernobyl. Is the patchwork of low-dose radiation actually promoting the health of populations of wolves, wild boar and other animals? In particular, the herds of Przewalski's horses that are the descendants of a tiny pool of equine diversity might have a lot to gain from a boost in their immune systems, antioxidants and DNA repair. Once a lethal influence, might the fallout from Chernobyl now provide a necessary shuffling of the genetic deck? Even if there is an increase in cancer in these animals, would that really impact their reproductive success or survival? Unlike humans, most wild animals don't live long after they've passed their prime reproductive age. The age-related diseases are such a problem for our species because we spend a long time post-reproductively ageing. A mutation that might lead to cancer later in life is of little significance to a young horse that might soon be trampled by a rival or eaten by a wolf. In a particularly harsh winter, he may starve to death or fall into an ice-covered lake. A cut on his flank can easily become infected and bacteria kill him in his prime. When seen from this non-human perspective, the exclusion zone around Chernobyl (once called 'the death zone'), offering a bit of ionizing radiation that might boost the immune system and increase DNA repair and antioxidant release, starts to sound more like a place of rejuvenation.

Since the Przewalski's horses never experienced the high doses of radionuclides that would have degraded in the first few months (iodine) or years (caesium), it could be a case of being in the right place at the right time. Seen this way, I understand why Orizaola is so dedicated to sampling the faeces of these animals. 'They're the most interesting species,' he tells me, later adding that the study of dogs in Chernobyl might make headlines but has several fundamental flaws, including the fact these so-called stray dogs are still fed and cared for by people who remain in the exclusion zone. 'But now we have to wait.'

We needn't visit exclusion zones or ponder tardigrades flying through space to marvel at life's ability to withstand radiation. Arguably the most significant discovery in radiobiology came from a can of minced beef in Oregon in 1954. As was standard practice at the time, tinned foods were sterilized with radioactive cobalt (60-Co). But scientists writing in the journal *Agricultural and Food Sciences Biology* in 1956 found that the meat wasn't completely sterile. There was one species of bacteria in the meat: *Deinococcus radiodurans*, a microbe that was often linked together into groups of four and had a distinctive pinkish hue. As with the Mariana snailfish, the toughest microbes on Earth are candy floss-coloured.

Similar to heat-tolerant ants and crucian carp, *D. radiodurans* is a subordinate member of microbial diversity. Usually found in very low numbers, a blast of ionizing radiation kills off its more dominant neighbours and allows this radiation-tolerant microbe to multiply without competition. A 400-gram lump of minced beef is their world to conquer. Since this discovery in 1956, *D. radiodurans* has become one of the best-studied organisms, pored over for any insights into how to survive levels of radiation that are 1,000 times that considered lethal for a human. It begins with the microbe's cell wall, that thick covering surrounding the more flexible membrane. Just as a roof is made up of slate, waterproof membrane, timber, insulation and plasterboard, the cell wall of *D. radiodurans* is multi-layered, providing a filtration system against the most explosive radionuclides. Inside the cell, it has packed its internal chemistry with manganese, an element that acts as a metallic armour around its proteins – those overzealous groundskeepers that can repair any damage to DNA and other protein complexes. A plethora of antioxidants mop up the inevitable reactive oxygen species.

Even compared to tardigrades, *D. radiodurans* is in a league of its own when it comes to radiation resistance. Scientists have even written a coded message into its microbes kept in the laboratory, a strip of DNA that might remain decipherable after a nuclear apocalypse.

And yet radiation resistance isn't difficult to evolve. In 2009, researchers from the University of Wisconsin-Madison and Louisiana State University grew a population of *E. coli*, a common species of bacteria that is radiation-sensitive, and forced it into a resilient microbe comparable to *D. radiodurans*. Their trick was nothing more than speeding up the process of natural selection. Exposing an initial population of lab-grown *E. coli*, the group, led by Michael Cox and John Battista, exposed the bacteria to levels of radiation that killed off 99 per cent of the cells, leaving only a few that could naturally grow in such extreme conditions. Generation after generation, the level of radiation was increased and the same 1 per cent of bacteria selected as the ancestors for the next population. With a generation time of a few hours, evolution was forced into hyperdrive. After just 20 generations, this once radiation-sensitive species could endure 3,000 Gray, a level not far off the record of 5,000 Gray that *D. radiodurans* can survive.

A few years after this study, a PhD student from Wisconsin-Madison, Rose Byrne, sequenced the genomes of these radiation-resistant *E. coli* and found that the changes were surprisingly few and remarkably simple: just three mutations in genes involved in DNA repair were largely responsible for the remarkable increase in radiation resistance. As Byrne wrote in the journal *eLife* in 2014, DNA repair systems are 'biologically malleable'. A rather over-enthusiastic journalist writing in *Scientific American* stated that this discovery could lead to radiation-resistant humans. (Instead of killing 99 per cent of wannabe astronauts over 20 generations, we might prefer to use genetic modification as used in gene therapy.)

Thought to be a minor component of soil microflora, *D. radiodurans* will never experience the level of radiation it has been exposed to in food preservation facilities and the laboratory. So why does this particular microbe have such an innate ability to survive an extreme environment it will never find itself in? As with the tardigrades, the answer is water – or a lack of it. A study found that microbes sampled from the Sonoran Desert in the US were also resistant to radiation. Anhydrobiosis, a life without water, prepares an organism for the worst case scenario: the evolution of a hyper-intelligent, naked primate that discovers the power of an unstable atom.

A day after we met outside Burgos to collect horse poo, I met Germán Orizaola and his colleague and wife Ana Elisa Valdés at their laboratory at the University of Oviedo, a blanket of grey cloud hanging above this campus south of the city. Unlike when we were outside in the open enclosures of Paleolítico Vivo, the smell of the faeces seems to add a density to the air, like sticking your head into a compost heap. Overlooking a central glass chamber of cacti and metre-tall snake plants, Valdés takes two-gram samples from inside each ball of poo and puts them into a test tube containing a buffer that protects the DNA before sequencing. Other balls will be left for two, four, seven, 14 and, finally, 28 days, each time testing whether DNA remains viable as the poo dries. By leaving stool samples out for various lengths of time, she can then analyse each and see when the amount of information needed is lost, either through contamination or degradation. This is important because, unlike in the prehistoric park, the Przewalski's horses in Chernobyl avoid humans as much as possible. (The earliest attempts of Western museums to catch these animals for their collection, it has been written, 'proved difficult... because they were too shy and fast'.) To pick up their poo won't be a matter of waiting, as it is in Spain, but also finding it. If Orizaola returns to Chernobyl, he will have to know where the horses spent the night – usually in a forested area – and picking the freshest dung he can see.

Next to Valdés, sitting on the same table and dressed in a casual polo neck shirt, Orizaola sieves poo samples through a tea strainer and looks for parasites under a digital microscope. They are mostly the eggs of worms and other intestinal bugs, each a control against what the horses in Chernobyl will contain. I ask them about their relationship, their children, and where they met. It was in Sweden, Valdés tells me, at the University of Uppsala. She was working on plant genetics while Orizaola studied Arctic frogs. Their interests, excepting the fact that they are biological, couldn't have been more different. They wanted to return to Spain where they both grew up to raise their children. Without saying a word, I wonder whether they ever expected they'd be sitting next to each other, poring over the faeces of an endangered horse.

As I leave their laboratory that day, one particular comment from Orizaola sticks with me: radiation disappears by itself. It may take weeks or years, but a radioactive particle will decay. 'It is a self-cleaning sort of pollution,' he says. Out of all the catastrophes caused by humans, Chernobyl pales in comparison to our dependence on fossil fuels, our addiction to plastics, and our use of inorganic molecules that are passed along food chains without ever breaking down. Our footprint grows and rarely dissipates. Radioactive iodine disappeared in the summer of 1986, a couple of months of intense ionizing radiation and then, poof, gone. There may be no precise date to pin down our impact on the natural world, no 26 April 1986 of radiation science, but it is the gradual emissions that will make the deepest impression. Our impact on the natural world is so pervasive and multi-faceted that James Lovelock – the proponent of the idea that the Earth is self-regulating, like an enormous organism – said that we should place radioactive waste in tropical forests to exclude humans and protect the animals that live there.

Epilogue

It's ironic that the greatest threat to life on Earth came from one of its own residents. Polluting the atmosphere with a novel gas, these wasteful organisms created an atmosphere inhospitable to nearly all other residents. It was a combustible gas, a molecule that hadn't been seen in such high quantities before. This moment in Earth's history has been called 'the greatest pollution crisis the earth has ever endured'.

But this is not today – the source of this pollution is not the burning of fossil fuels. Roughly two billion years ago, bacteria that lived in our oceans evolved an entirely new way of sustaining their cells: splitting water using the sun's rays. Absorbing carbon dioxide from the atmosphere, these so-called cyanobacteria discovered the power of photosynthesis. The carbon from CO_2 was paired with the hydrogen from water to build sugars and other carbon compounds. The oxygen stuck together into O_2 ('dioxygen'), a waste product. With previous concentrations of oxygen estimated to be 0.0001 per cent, the atmosphere became suffused with this previously insignificant gas. A world once dominated by anaerobic microbes was suffocated. As Earth started to breathe (as we would define it), the life it once sustained was destroyed. 'When exposed to oxygen and light, the tissues of these unadapted organisms are instantly destroyed,' wrote Lynn Margulis and Dorion Sagan in their book *Microcosmos*. '[B]y about 2,000 million years ago, the available passive reactants in the world had been used up and oxygen accumulated rapidly in the air, precipitating a catastrophe of global magnitude.' Even though life was purely microbial, it has been called the greatest extinction event in the history of the Earth.

In the wake of this catastrophe, life flourished. To protect themselves from this toxic gas, microbes joined forces – fashioning symbioses that were more stable and more productive than life spent alone. The evolution of complex life – all fungi, plants and animals – occurred in an oxygenated ocean. The roots of our tree of life reach into the biggest shift in biological dominance the world has ever seen. 'The evolution of oxygen-producing cyanobacteria was arguably *the* significant event in the history of life after the evolution of life itself,' writes Martha Sosa Torres and her colleagues in a book chapter called 'The Magic of Dioxygen'.

Whether this extinction event ever happened is still debated. Perhaps the microbes already on Earth had evolved an innate resilience to oxygen after billions of years of coping with the harmful, DNA-splitting effects of radiation. There is very limited fossil evidence from this ancient period of Earth's history. But even if it wasn't the 'Oxygen Holocaust' that Margulis and Sagan called it, this was an unimaginable shift in life on Earth. And it came from a group of microbes that would have spread throughout the Earth's oceans, soaking up water and pumping out oxygen wherever there was sufficient sunlight.

The Great Oxidation event was a prelude to other mass extinction events that were caused by life's restless pursuit for innovation. At the end of the Ordovician Period, a time when animal life was largely limited to the oceans and the largest predators were sea scorpions the size of dining tables, a global glaciation event led to the first of the big five mass extinctions. The cause? There was no asteroid or continent-splitting volcanic activity. Roughly 440 million years ago, trees evolved on land, growing deep roots and, eventually, reproducing with hardy seeds that could reach into the dry, inhospitable interior of the supercontinent, Pangaea. As they spread, these giant plants soaked up carbon dioxide from the atmosphere, turning a greenhouse world into an ice world. As Peter Brannen writes in his 2017 book *The Ends of the World*, 'Although trees today are seen as beneficent givers of life… these first forests on the planet might have heralded the end-times'. An estimated 85 per cent of species died.

The next iteration of life-killing-life would emerge much later – after the continental volcanoes of the Permian extinction and the asteroid and volcanoes of the late Cretaceous – when, some 100,000 years ago, a species of upright-walking ape evolved another unprecedented, world-changing adaptation: intelligence. While other animals had the cognitive powers to adapt and learn from their changing world, *Homo sapiens* became the ultimate expression of this evolutionary trend, surpassing any primate, bird or cephalopod that had come before. And, like the first photosynthetic cyanobacteria before them, humans discovered a powerful source of energy that was just waiting to be utilized: hydrocarbons. The fossilized remains of carbon-rich life, gas, coal and oil powered their movement, their communication and their dominion over life on Earth. This moment, only two centuries ago, would release another invisible and odourless gas into the atmosphere: carbon dioxide. The climate change we now see around us in extreme heat waves, uncontrollable fires and superpowered hurricanes all stem from this moment in our history. We understand it in exquisite detail and can even predict the future. By 2100, one prediction estimates, millions of people will die every year from the heat. But even when our own mortality is threatened (never mind the animals we share this world with), we are still sluggish to stop it.

Personally, to have a precedent – cyanobacteria and trees before humanity – makes our impacts on the world easier for me to reckon with. Evolution is married to extinction. But the Great Oxidation event, the Ordovician extinction and modern climate change aren't perfectly comparable. First, cyanobacteria and trees were oblivious to their impact on the world around them. We, on the other hand, are perfectly aware of our impact, and have been for many decades. As we pollute the atmosphere and make the Earth warmer, more fiery and chaotic, we do so consciously and largely without remorse. Second, photosynthesis – whether in cyanobacteria or the first trees – requires the inhalation of carbon dioxide and the exhalation of oxygen. It is an integral part of their biology. We don't *need* fossil fuels to survive. And this means that we can change. While cyanobacteria, volcanic eruptions and asteroid strikes kill unrepentantly, unconsciously, we are the first conjurers of mass extinction events that can modify our actions to help nature heal.

We are far from achieving the ultimate goal of net zero carbon emissions, a global population that is fed, powered and transported without any additional carbon dioxide released into the atmosphere. The modern rate of carbon emissions is unprecedented (even compared to the Great Dying of the Permian extinction), and our capacity to change is not yet fit for the challenge ahead. But behind the doom and gloom, away from the unscientific idealism of Just Stop Oil, there are incredible transitions already under way. Just as life has evolved to harness novel sources of energy – radiation, sunlight, sulphur – we are transitioning away from fossil fuels to forms of renewable electricity. Once thought to be too expensive to become a worthy successor to coal, oil and gas, solar and wind are now the most affordable options for over 80 per cent of the world. Between 2010 and 2020 alone, the costs of the photovoltaic cells used in solar panels fell by 85 per cent.

To harness the power of sunlight and wind is one thing. To recreate – and utilize – the most powerful explosions in the universe is another. On 5 December 2023, researchers at the Lawrence Livermore National Laboratory, 160 kilometres east of San Francisco, fired 192 high-powered lasers onto a sphere of hydrogen atoms the size of a poppy seed, heating them to temperatures found in the centre of the sun. The resulting fusion reaction had only previously been created on Earth with the release of hydrogen bombs in the middle of the twentieth century. This time, the self-sustaining reaction could be initiated in a controlled setting, a starting line for the race towards everlasting energy. From Robert Oppenheimer's 'Now I am become death, the destroyer of worlds', fusion technology has become a potential saviour of Earth.

An exciting possibility, fusion technology on any significant scale is likely to take decades to become a commercial reality. The cost of powering hundreds of giant lasers, the production of thousands of hydrogen-containing capsules made from diamond, and the inefficiency of current technology (two units of energy in, three units out) all require huge investment. Kim Budil, director of the Lawrence Livermore National Laboratory, likens the first fusion reaction to the Wright brothers' first flight. In December 1903, 120 years before her lab's fusion reaction, the *Wright Flyer* was airborne for 12 seconds and travelled 40 metres, less than half a football field. The only difference is that modern aeroplanes have only added to the carbon emissions, whereas fusion technology would reduce them.

While our own ingenuity lessens our dependency on fossil fuels, there are ways we can dramatically reduce our carbon footprint without any technological revolutions. Agriculture contributes between 25 and 30 per cent of the world's carbon emissions, far more than air travel, which adds less than 4 per cent. A study into 55,000 people living in the UK, published in the journal *Nature Food* in 2023, found that a plant-based diet can reduce a person's carbon footprint by 75 per cent compared to 'high-meat eaters' who eat over 100 grams of meat a day. In addition to reducing greenhouse gas emissions, similar reductions were found in land use, water consumption and water pollution – the eutrophication that leads to 'dead zones' in our lakes and oceans. Not limiting themselves to cold data that can be easy to ignore, the authors clearly saw a need to make a more conversational plea for change: 'the relationship between environmental impact and animal-based food consumption is clear,' they wrote, 'and should prompt the reduction of the latter'.

While this fight for greener economies and renewable energy goes on, scientists and conservationists are busily trying to keep their species and the ecosystems from falling into the abyss, helping them build and maintain resilience to a more inhospitable world that is dawning. Around the world, branching corals are being bred, and hybridized, to produce more resilient reefs that can endure warmer, more acidic oceans. Trees are being moved northwards as climate shifts outpace their ability to set seed and 'migrate' over generations. And while individual species – especially those such as elkhorn coral – can be foundational to an entire ecosystem, it is also true that biodiversity is the key to long-term resilience in a changing world. The more species that live together, the better they can deal with change, whether it is gradual or immediate. Biodiversity is a natural buffer against the inevitability of climate change. This is why Marine Protected Areas in the oceans or National Parks on land are so important; they can't change the climate but they can preserve a varied piece of nature, an assemblage of animals, plants, fungi and microbes that might stand a fighting chance as their surroundings shift.

The same is true for geology. The more diversity of the rocks and soil in an area, the more resilient it is to changing climates. In the United States, maintaining the connections between these 'resilient hotspots' has become a priority, enabling species to move between different microclimates and find refuge for years to come. Even when their former range becomes inhospitable, there are places nearby – shaped by geology, biodiversity and climate – that will still feel like home. As a 2023 study published in *Proceedings of the National Academy of Sciences* states, 'Rather than solely reacting to near-term drivers of species and habitat loss, this network [of geophysical diversity] can help conservationists envision how to allow nature to adjust to a rapidly changing earth, with a goal of sustaining a dynamic, diverse, and adaptive natural world.' This broad, long-term approach to conservation has been called 'conserving nature's stage'.

Whether conservation strategies focus on keystone species, biodiversity or geology, there will undoubtedly be winners and losers. Species will go extinct; subordinate – often weedy – taxa may rise to dominance. But life will persist.

I often turn to life itself when all hope seems lost. What the animals, plants and fungi in the previous pages have taught me is that life has a resilience that surpasses any calamity. Tardigrades would only die if the oceans boiled away into space. Forams thrive in the dead zones we are creating in our oceans. Even polar bears, those images of impending doom, are adapting and finding new places to thrive as the ice melts.

There will be heartbreak; incredible, beautiful and unique life forms will be lost forever. Just as overhunting exterminated the dodo and the Tasmanian tiger, climate change will claim its own victims as habitats become too warm, too anoxic or simply too chaotic for even the hardiest species to endure.

Even in the worst case scenario of a sixth mass extinction, the history of life on Earth shows that recovery isn't just possible but is a necessary part of new forms of life. Seen through the lens of deep time, extinction is a natural part of a living planet. The Great Dying that killed off 96 per cent of all marine life ushered in a fierce world of predator and prey that we call home today. The asteroid that wiped out the dinosaurs paved the way for the reign of mammals.

Life, once it has emerged on a planet, is very hard to destroy. Looking forward millions of years into the future, I wonder whether ecosystems will be defined not by competitiveness or predation but by cooperation and harmony, a counterweight to the human actions that proved so devastating.

Since I started researching this book, my faith in the ingenuity and resilience of Earth's inhabitants has only deepened. Wherever there is a source of energy to be utilized, evolution will find a way to harness it. The radioactive splitting of water into hydrogen, sulphur emitted from hydrothermal vents, sunlight: the foundations of ecosystems are commonplace. Looking further afield, there are planetary bodies in our own solar system that are probably habitable, even if they aren't inhabited. The discovery that life can emerge without the influence of the sun's rays expands the potential range of habitable planets in the universe. The so-called Goldilocks zone, a region in a solar system where light and temperature are just right for water to remain liquid, is now just a slither of the search image for extraterrestrial life. Just as the moons of Saturn and Jupiter are geologically active due to the gravitational jostling of their neighbours, oceans can form even in the most distant regions of a solar system. When seen through the trillions of solar systems in the universe (all forming a part of the billions of galaxies), there are countless moons and planets that could harbour life, even in darkness.

The rotation of our planet keeps life cycling through days and nights on a regular and repeatable cycle. Knowing that life can remain dormant for months on our planet – turtles in anoxic ponds, tardigrades in dried-up moss, frogs in blocks of ice – I can imagine a planet, somewhere in the limitless universe, that doesn't rotate but each orbit around its sun brings a day as long as our seasons, each side facing the sunlight for half the year and then facing away for the other half. To survive, the creatures here consume the rich foodstuffs like a polar bear feasting on seal blubber, and then prepare for the coming harshness of a celestial winter. Perhaps they burrow into the earth like a ground squirrel or just shrivel up into a tardigrade-like tun, immobile and impenetrable.

Next to my laptop and stacks of notes on my desk is a glass bowl with some dried-up dirt inside, the sample in which I spotted my first wild tardigrade. Their world disappeared as the water evaporated. They will revive from their tun state, their brazen 'chemical indifference' to all stress, if I add a drop of water. My actions – leaving the glass jar to dry out – are of no real consequence to their survival. Tardigrades, as a group, have survived all five mass extinctions on Earth, their ability to endure the most hostile influences seeing them through the hard times like a seed buried deep in permafrost. I planned to add water to the bowl before writing this epilogue, adding an account of their emergence from suspended animation. In the end, I decided to leave them a little longer. This bowl represents something much more powerful, a tangible microcosm of defiance, a reminder of resilience that fits in the palm of my hand.

Acknowledgements

I have had the pleasure of meeting over a hundred scientists while reporting the stories in this book. Whether in person, at their laboratories or out in the field, each conversation has shaped the structure and content of the chapters. Some sources have been most generous with their time, spending a morning or afternoon explaining their science, answering follow-up questions over email. As such, my deepest thanks are reserved for Mackenzie Gerringer, Johanna Weston, Les Buck, Matt Pamenter, Germán Orizaola, Pablo Burraco, Xim Cerdá, Ricardo Amils, Ellis Moloney, Brian Lewarne and Steve D'Hondt.

As well as offering thanks, I would like to add an apology. A book-sized project requires pruning if it is to grow into its most accessible form. A few people I have met or spoken to haven't been included in the final text, but I hope there is some consolation in knowing that every conversation has shaped the book's content.

For taking the book from a rough proposal into a

clean manuscript, there are several people to thank: my incredible agent, Carrie Plitt at Felicity Bryan Associates; my editors Jessica Yao at W.W. Norton and Drummond Moir at Atlantic Books; Tamsin Shelton for a thorough and thoughtful copy-edit; and James Nightingale and Poppy Hampson, who first bought the rights for the book in the UK. I would like to thank those who read through the first draft on the unacceptably tight deadline I set to ensure I got most of the science right: Ellis Moloney, Andrew Derocher, Julia Sigwart, Barbara Sherwood Lollar, Kristin Laidre, Germán Orizaola, Arturo Casadevall, Melody Clark, Dylan Schwilk, Matt Pamenter, Sabrina Elkassas, Chris Guglielmo, Ekaterina Dadachova, Lloyd Peck, Mackenzie Gerringer, Les Buck, Paul Yancey, Almut Kelber, Frank van Breukelen, Xim Cerdá, Eric Warrant, Joan Bernhard, Don Larson, Brian Barnes, Cindy Morris and Johanna Weston.

Finally, to my family. Pregnant with twins and caring for a three-year-old, my partner Lucy has braved an extreme environment worthy of this book's chapters while I travelled to universities, laboratories and field sites across Europe and North America. Although I tried to streamline my reporting trips, I was often away for a week or more. Her strength and ability to endure near-constant sickness and levels of discomfort that I can't imagine are my daily reminder of life's resilience (along with my desktop tardigrades). Bringing our beautiful boys into the world has made us both stronger, although it may not feel like it as we endure their highly orchestrated tag-team of sleeplessness. And to my incredible daughter Nieve, as determined as any bar-headed goose braving the Himalayas, I am very lucky to spend my time watching you grow. I have no doubt that you will find your own special niche in this world.

This book has been written in the dark hours of the night and on those days when child-minders and grandparents released the pressure of parenting for a few hours. After several rounds of edits, copy-edits and review from the scientists I've spoken to, any errors are down to my sleepy brain. It has been a joy to research and write, an escape from daily stresses and the constant barrage of depressing news. I hope it has been a joy to read.

Notes

Introduction

1 https://www.youtube.com/watch?v=kux1j1ccsgg&ab_channel=JourneytotheMicrocosmos

Chapter 1: Dry Hard - Water

1 The original laboratory stock of *Hypsibius examplaris* was collected from a pond in 1987 by Bob McNuff, an amateur tardigrade enthusiast who lived just outside of Manchester, northwest England, spending his spare time in his garden laboratory. While neighbours kept lawn mowers, compost and stacks of spiderwebbed plant pots in their sheds, McNuff's space was clean and dedicated to the breeding of *H. exemplaris*. Before McNuff went into hospice care, he made sure that his stock of water bears found a new home. As a tardigrade researcher told me, 'He made sure to send his tardigrade stocks to Carolina Biological [a laboratory products company], so that there'd always be a commercial supply. Remarkable! I'm not sure it'd be on my bucket list if I were entering a hospice.'

2 Tardigrade taxonomists estimate that there are at least twice that number.

3 Not as famous as tardigrades, rotifers are incredible animals. As a group, they haven't had sex for millions of years. Producing males hasn't been a part of their history. Multiplying by asexual means, they have increased their genetic diversity – and therefore been able to evolve – by sucking up the genetic material of their neighbours, both microbe and animal. Such 'horizontal gene transfer' was once thought to be impossible. In fact, it is the key to the sexless rotifers' success. As one study in 2014 puts it, rotifers 'have been able to survive and diversify for millions of years without sex'. (Hespeels et al., 2014.)

4 We start to feel dizzy and convulse at 10 per cent water loss; 20 per cent is lethal.

5 The reason sea monkeys could be sold as pets was completely down to their tolerance to desiccation. Dried out, they can be transported to pet shops and people's houses. Add water, and they emerge and swim. Add to this familiar animal the even more familiar yeast, and these single-celled fungi can remain in a desiccated state for months, a trick that allows them to survive the dry season in wine-producing regions such as California.

6 The same happens in humans. If we lose 15 per cent of our body's water, most of which is drawn from the plasma and turned into sweat, we experience a lethal explosive heat rise. The thing that keeps us cool – evaporation of sweat – also kills us through dehydration.

7 Whether this is an adaptation or an unavoidable reaction to a lack of water is still debated. For our purposes, it's enough to know that it occurs and let the academics iron out the semantics.

8 'Giant' is a relative term. This species, the largest of the 20 species of kangaroo rats, is still small enough to sit on your hand and have room to turn around.

Chapter 2: Breathtaking (Oxygen)

1 A geological epoch defined by continental ice sheets that began 2.58 million years ago.

2 Technically 'glycogen', a molecule composed of long chains of glucose.
3 Lactic acid is the textbook outcome of anaerobic metabolism. Shockingly, as I researched this book, I realized that it was wrong; we don't produce lactic acid. Not even when we have exercised. It just doesn't happen. The truth is only slightly different, but to a biochemist, slight differences are the most important. Without oxygen, a muscle cell actually releases lactate, a negatively charged particle that doesn't acidify our cells. The acid comes from elsewhere, the breakdown of ATP, the energy currency of a cell, into its component parts: ADP and a proton (H+). It is the build-up of these protons that generates acidity. The pH scale for the acidity (or alkalinity) of a solution is, literally, the '**p**ower of **H**ydrogen'. The more H+ ions, the more acidic a solution is. Blood is a neutral pH of 7.4, for example, and anaerobic metabolism disrupts this neutrality. But it's easier to just say lactic acid, isn't it?
4 These are the adults. The life cycle of these animals is incredibly complex (with several larval *and* post-larval stages) and still being unravelled. One larval form, known as the Higgins larva, has a flipper or spiny toe that it uses to move through the waters among the plankton.
5 Technically 'organelles', but chloroplasts, and mitochondria, act like organs of the body but contained within a cell.
6 Anoxic oceans may be called 'dead zones' today but they were the norm for much of Earth's history. While photosynthesis evolved around two billion years ago and began to flush the atmosphere and shallow seas with oxygen, those places that were below where the light touches (around 200 metres down) remained anoxic, or extremely hypoxic. It wasn't until 500 million years ago that the deep sea started to welcome aerobic life into its waters. Seen through this lens, the current trend of deoxygenation and the animal life that dwells in anoxia seem slightly less alien. A Loriciferan growing and reproducing at the bottom of a stagnant basin could even be a glimpse into an

ancestral state for all animals, a homecoming to a time when anoxia was the norm.

7 This geographical limit isn't where temperatures become too cold. At more northerly climes, the winters are just a little bit too long, the spring and summer too short, for the turtles to emerge from their overwintering ponds and mate with another turtle. Slow and steady might win in a race against a cocky and impetuous rabbit, but it doesn't fare so well against the inevitable freeze of a Canadian winter. As climate change warms our winters into ever-shorter periods of freezing, these turtles are likely to move – slowly – further north, breaking new ground and perhaps, in the not too distant future, breed within the Arctic Circle.

8 First described in the late nineteenth century, scientific interest in naked mole-rats really began in earnest in 1981. That year, Jenny Jarvis, a PhD student at the University of Cape Town, discovered their secret of eusociality. Not only was there a breeding female in this species, but there were several types of workers or 'castes'. Somc were excavators. Others looked after the young. And then there were the large males who did little but lie about and occasionally mate with the queen. In a later paper, Jarvis described a particular caste known as the 'volcanoer', another large individual whose job is to push the dirt and debris from excavation out of the tunnels and into the world above.

9 One female caught in the wild gave birth to 900 pups in 11 years.

10 There are many misunderstandings when it comes to naked mole-rats, so much so that dozens of researchers wrote a review in 2021 that identified 28 'myths' about naked mole-rats. Although they have lost receptors for certain pains, they will squeal if you pinch them. While they have low rates of cancer, they aren't immune to it. They do age. From pain to cancer, thermoregulation to hypoxia, there is a tendency to add hyperbole to anything *Heterocephalus*. (Buffenstein et al., 2022.)

Chapter 3: Fasting & Furious (Food)

1 The length of a European starling is between 19 and 23 centimetres while the length of a common poorwill is between 19 and 21 centimetres.

2 A belief that is still written into their family's Latin name: Caprimulgida*e*.

3 They are also defined by the presence of a scrotum in males.

4 While the smoking gun for this mass extinction is still debated, it is likely that the release of enormous quantities of lava increased global temperatures by 5°C (over millions of years) and poisoned the oceans with hydrogen sulphide, sapping sea life of its oxygen.

5 0.73 square kilometres.

6 Brown rats have also been introduced but they didn't evolve on the island.

7 Most commonly two cubs but sometimes one or three (and very rarely four).

8 The largest polar bear was a male that weighed 1,002 kilograms (just over a metric tonne). A large brown bear, or grizzly, might weigh 750 kilograms. The largest non-bear carnivore is the tiger, at 300 kilograms.

9 Welcome to the family Ursidae: black bear, Asiatic black bear, brown bear, panda bear, sloth bear, sun bear, spectacled bear and polar bear. While there are many cute fashion dogs called 'Bear' they are not related.

10 This description is based on Sherman, a rhesus monkey that lived in the National Institute on Aging in Maryland.

11 Magnesium and potassium are also depleted and lead to dysfunction in firing neurons and contracting muscle fibres.

12 Birds were once thought to stay within their summer ranges but just play a very convincing game of hide and seek.

13 The Arctic tern could also be called the Antarctic tern.

14 In recent years, 30 per cent of the mudflats of the Yellow Sea have been lost to land reclamation projects or pollution. Since 100

per cent of a subspecies of bar-tailed godwits (*Limosa lapponica menzbieri*) use the Yellow Sea as a stopover habitat, it is unsurprising that their populations have been declining in recent years.
15 The moon is 383,400 kilometres from Earth.

Chapter 4: Supercool Animals (Freezing)

1 This can partly be explained by the lower metabolic rate of the icefish, thereby making their cardiovascular system have a greater proportion of their energy budget since everything else takes up so little.
2 Ocean Floor Observation and Bathymetry System.
3 In McMurdo Sound, these delicate filter-feeders cover 55 per cent of the bottom.
4 Not true for all clouds – only when temperatures are below 0°C.
5 Not all insects supercool. There are many examples of insects using sugars like glucose and freezing – wood frog-style – into a controlled block of ice. The caterpillars of the Arctic woolly bear moth remain frozen for 11 months of the year, only emerging to feed for a few weeks in the summer. (Kukal, Serianni & Duman, 1988.) With such a short window of growth, they can take over a decade to metamorphose into their winged adult stage. A moth fluttering over the grassy tundra of Canada's Ellesmere Island, a place of reindeer-hunting polar bears, might be 14 years old.
6 Cryopreservation shouldn't be confused with cryology, the field of pseudoscience that freezes human bodies after brain death in the hope that future medical technologies could cure their disease. Despite some undeniably moving accounts of people wanting to return to society at a time when their terminal cancer can be treated, such rebirth after death is primarily a means of coping with death, a ray of hope that there may be an afterlife.
7 'Clinically acceptable survival' was the term at this stage. At 96 hours, only 58 per cent of organs were clinically useful (i.e., for transplantation).

8 A human liver is 1.5 kilograms. A rat's just 5 grams.
9 Darkness can be longer than two months. At the South Pole, it can last for six months.
10 With penguin colonies comprising hundreds or even thousands of breeding pairs, there can be a lot of poo, a stain on the icy surface that can be seen from space. In photographs published in the journal *Antarctic Science* in 2023, the presence of a colony of penguins is a subtle – but distinct – freckle on the otherwise colourless photograph.

Chapter 5: Highs and Lows (Pressure)

1 A European rabbit weighs 1–2.5 kilograms; a bar-headed goose weighs 2–3 kilograms.
2 The Andes average 4,000 metres.
3 This figure is around 1 per cent in high-altitude populations in Tibet, partly because they don't respond to altitude by producing haemoglobin. In fact, their ability to live happily at altitude might be related to a genetic mutation that Tibetan people inherited from Denisovans, an extinct species of hominin that humans bred with some 40,000 years ago. (Huerta-Sánchez et al., 2014.)
4 The highest resident mammal, first discovered in 2019, is the yellow-rumped leaf-eared mouse, a species that lives at 6,200 metres in the Andes.
5 Swan and Hawkes… It's not a prerequisite to have an avian surname to study high-altitude birds, but maybe it helps.
6 Cold War and modern Russian atrocities aside, the Soviet terminology does seem to be more suited to these depths. The term 'hadal' suggests another world, one beneath the ocean and not connected to it. Ultra-abyssal, meanwhile, is an extension of the ocean as we know it. It might not reflect the unique qualities of the animals that live there but it is nonetheless more fluid in our appreciation of the different realms of this planet.
7 This isn't completely wrong; fish like this do live at depth. Anglerfishes are the classic example, ambush predators that use

a glowing orb on a thin rod of flesh to attract prey towards their bear-trap mouths. But these fish are found in the 'twilight zone', a place where some light still penetrates.

8 Also renamed grenadiers by fishermen and wholesalers who, correctly, thought that rat-tails wasn't an appetizing label to put on the day's catch.

9 Gerringer and Linley chose the species name *swirei* in honour of Swire Deep, the 8,000 metre part of the trench discovered by the *Challenger* expedition in 1875.

10 Another contender for deepest fish was a cusk eel caught at 7,965 metres in the Puerto Rico Trench in 1970. (Linley et al., 2016, p. 5.) However, this is now believed to have been caught as the open net was brought to the surface and not at the maximum depth. And the so-called Trieste flatfish, which was observed from the pioneering *Trieste* submersible dropped to the bottom of the Mariana Trench in 1962, was more likely a sea cucumber, an oval-shaped species that is about 30 centimetres long and could be mistaken for a flatfish.(Jamieson & Yancey, 2012; Wolff, 1961.)

11 A fat molecule looks a bit like a lollipop. There's a bulbous end (an atom of phosphorus) and a stick-like tail (the lipid chain). In a membrane, they form two layers of lollipops, their tails touching and their round ends either facing inside the cell or outside the cell. It's called a phospholipid bi-layer. It looks a bit like this:

```
0 0 0 0       –    phosphorous end (water-loving)
|| || || ||   –    lipid tail (water-hating)
|| || || ||   –    lipid tail (water-hating)
0 0 0 0       –    phosphorous end (water-loving)
```

In most animals, the lipid tails are straight, lining up to their neighbour in perfect parallel. But they can also have kinks in their structure, as if the lollipop stick has been chewed and bent (from this '||' to this '>>'). This small tweak stops the membrane from being packed together too tightly. The fatty tails don't line up neatly, and space is maintained for movement.

12 When broken down by bacteria, TMAO loses its O (oxygen) and becomes TMA, the fishy smell we associate with the ocean or out of date seafood.
13 Whether TMAO or salt, i.e., sodium chloride, the amount of 'stuff' that is held in water changes its osmolarity. Measured in milli-osmoles, the internal chemistry of fish is kept at around 300–400 milli-osmoles. Seawater, meanwhile, with its dissolved sodium chloride, hovers between 1,000 and 1,100 milli-osmoles. To account for the difference, the gills and kidneys of fish have become very good at removing salt from the seawater that they drink. They invest a lot of metabolic energy in this process, and they have to drink often in order to account for the surrounding seawater that is constantly dehydrating their cells. As fish produce more TMAO, however, they edge closer and closer to being in harmony with their surroundings.
14 Each fish has six ear bones, and Gerringer selected the two largest for her estimates.

Chapter 6: Life in the Furnace (Heat)

1 Saharan silver ant: 53.6°C but with a range either side of 0.8°C, i.e., between 52.8°C and 54.4°C.
2 The critical thermal maximum (or CTMax for short) is the temperature at which normal locomotion ceases, the point at which an individual wouldn't be able to escape from a 'thermal trap'. And the Saharan silver ant doesn't hold the title for the highest CTMax. In the hot shrublands of central Australia, the red honey ant (*Melophorus bagoti*) has a maximum temperature of 56.7°C, leading scientists who study this species to claim that 'it may never be too hot in central Australia for foraging by *M. bagoti*'. (Christian & Morton, 1992.)
3 Specifically, a scanning electron microscope, or SEM, a machine that fires electrons into a sample of tissue, and a detector underneath the sample builds up a picture based on where the electrons are deflected (or not).

4 This was actually the title of one of Rüdiger Wehner's papers, 'Desert ants on a thermal tightrope', published in *Nature* in 1992 with his wife Sibylle Wehner and a colleague from the University of Namibia, Alan Marsh.

5 As a study from 2003 states, 'Few of these deaths are recognisable clinically as being due to heat. Heat stress causes loss of salt and water in sweat, causing haemoconcentration, which in turn causes increases in coronary and cerebral thrombosis.' (Keatinge, 2003.)

6 To a physicist, heat is defined as the *transfer* of thermal energy from one object to another, whether it's a radiator emitting heat into a room or the sun warming the Earth's surface. The temperature of an object is more accurately the 'thermal energy' it contains. I hope the *thermal energy* of any physicist reading this is slightly cooled with this explanation.

7 Although our bodies remain stable despite their changes in our environment, we release Hsps during a fever.

8 Still, a lot can happen over four billion years. And microbes are nothing but ever-changing. Bacteria can share genes like collectable cards, a process known as horizontal gene transfer that can give a neighbour (and not a descendant) the genetic tools to survive in a new environment. If the early microbes were cold-adapted and secondarily attained the love for hot environments, the reversal is blurred by the passing of time and this quirk of microbial inheritance.

9 With a wingspan of 70 centimetres.

Chapter 7: Ain't No Sunshine (Darkness)

1 Technically, this isn't quite true. The oxygen that these animals depend on is still tied to the photosynthesis of land plants and cyanobacteria in the oceans. But in terms of nutrition, they are essentially independent of sunlight.

2 The deepest black smokers currently known are situated in the Cayman Trough, 5,000 metres down in the Caribbean Sea. (Connelly et al., 2012.)

3 Only two to four times more ancient.
4 Its size, body pattern and fondness for honeysuckle make me pretty confident that this was an elephant hawkmoth. I asked my go-to moth specialists and they agreed that it was most likely.
5 The closest solar system to our own is Alpha Centauri, 4.3 light years away.

Chapter 8: Taste of a Poison Paradise (Radiation)

1 Once thought to be the last remaining wild horse, a recent analysis of ancient DNA found at Botai, northern Kazakhstan, suggests that Przewalski's might be the descendants of horses that were domesticated some 7,000 years ago. (Gaunitz et al., 2018.)
2 One banana contains 1 per cent of the radioactive exposure a person might receive from background radiation in a day. In other words, a tiny amount.
3 Other research groups argue that the opposite is true. Sieving through soil and surveying bird populations, researchers have found that higher levels of radiation correlate with lower biodiversity. Orizaola and other researchers say that these conclusions are based on too few samples and unnatural measurements of radiation levels – not incorporating the important component of time in their equations, for example. With these criticisms in mind, I tend to follow the replicated finding – high biodiversity in Chernobyl – rather than giving the same weight of discussion to a couple of researchers who consistently find the opposite.
4 Writing in *Nature* in 2003, two toxicologists described this as such a significant shift in our conception of how biological systems respond to radiation and heavy metals that it was similar to a change 'from a Soviet-style society to a western one'. (Calabrese & Baldwin, 2003, p. 692.)

References

Introduction

'a little puppy-shaped animal'... Slack, 1861.
'extraordinary force of resistance'... Mathews, 1938, p. 620.
'impossible as yet to state'... Ibid., pp. 620–21.
More recent studies have replicated... Goldstein, 2022, p. 188.
250 to 290 kilometres... Rebecchi et al., 2011, p. 98.
a few hardy members of *Milnesium tardigradum* survived... Jönsson et al., 2008.
'For complete sterilisation [of the planet]'... Sloan, Batista & Loeb, 2017.
tardigrades were impervious to the hotter, drier conditions... Vecchi et al., 2021.
'It is, of course, axiomatic'... Grinnell, 1917, p. 433.
'life's eternal dissatisfaction'... Eiseley, 1957.
'Those who contemplate'... Carson, 1962.
'Hope'... Solnit, 2005.

Chapter 1. Dry Hard (Water)

1,380 species of known tardigrades… Goldstein, 2022.
***H. exemplaris* and *Ramazzottius varieornatus*…** Ibid.
***Hypsibius* allows itself to be torn apart…** Clark-Hatchel et al., 2024.
***Ramazzottius*, meanwhile, protects itself from damage…** Chavez et al., 2019.
1,000 times that which would kill a human… Goldstein, 2022, p. 178.
nematode worms… Keilin, 1959.
a species of midge found in Central Africa… Cornette & Kikawada, 2011.
they remove up to 98 per cent of their body's water… Møbjerg & Neves, 2021, p. 1.
anhydrobiosis, or 'life without water'… Crowe, 2015.
See Note 3: **'millions of years without sex'…** Hespeels et al., 2014.
dispersed by the wind… Nelson, Bartels & Guil, 2019, p. 186.
'the microscopic animals are not at all troubled'… Murray, 1910.
cryoconite holes… Nelson, Bartels & Guil, 2019, p. 171.
chemical indifference… LePrince & Buitink, 2015, p. 372. Claude Bernhard, 1878.
'seems to provide animals'… Møbjerg & Neves, 2021, p. 1.
'escape in time'… Rebecchi, Boschetti & Nelson, 2020, p. 2790.
collected from the seabed… Nelson, Bartels & Guil, 2019, p. 165.
14,000 tardigrades in a square centimetre… Ibid., p. 79. Maximum of 14,000,000 per metre squared.
found on mountain peaks… Ibid., p. 170.
sucking discs on the ends of their legs… Ibid., p. 174.
***Bergtrollus dzimbowski*…** Dastych, 2011.
***Tanarctus bubulubus* has 16 to 20 balloon-like organs…** Jørgensen & Kristensen, 2001.
80 per cent of known tardigrade species are terrestrial… Nelson, Bartels & Guil, 2019, p. 172.
'The most relevant point'… Greven, 2019, p. 8.

these seemingly impossible reviviscences became a popular pastime… Keilin, 1959, p. 156: 'There is not at this day any professor, any amateur of natural history, particularly in Italy, who does not take pleasure in amusing himself, and gratifying the curiosity of his learned friends with these admirable resurrections.'

'cryptobiosis'… Ibid., p. 166.

'begin anew to actuate the same body'… Ibid., p. 153.

'partly extended specimen'… Jönsson & Bertolani, 2001, p. 122.

'tardigrades were later found crawling all over it'… Ibid.

ten years without water… Guidetti & Jönsson, 2002.

Sleeping Beauty 1 and Sleeping Beauty 2… Tsujimoto, Imura & Kanda, 2015.

trehalose… key to surviving desiccation… Westh & Ramløv, 1991.

(LEA) proteins… (SAHS) proteins… (CAHS) proteins… Goldstein, 2022.

330 known species of such 'resurrection plants'… Oliver et al., 2020.

recommence photosynthesis within a few hours… Ibid.

concertina-like… reduce its size by over 80 per cent… Ibid., p. 7.5.

exposing their undersides to the sunlight… Ibid., p. 7.10.

digest their chloroplasts… Ibid., p. 7.8.

same mechanism that other plants use in their seeds… Farrant & Hillhorst, 2022.

50 per cent of the world's plant-derived energy… Ibid., p. 84.

land that is too dry for agriculture… Dai, 2011.

a single shower during a thunderstorm… Henschel et al., 2018, p. 3.

'a stranded Octopus'… Cooper-Driver, 1994, p. 5.

'seedling in arrested development'… Ibid., p. 6.

leaf extending to more than six metres… Krüger et al., 2017.

***Welwitschia*'s roots don't go much deeper than a metre…** Henschel et al., 2018.

'a monster like *Welwitschia*'… Cooper-Driver, 1994, p. 2.

50,000 of these plants… Henschel et al., 2018, p. 4.

photosynthesis is incredibly efficient… Krüger et al., 2017.
'two leaves that cannot die'… https://www.kew.org/plants/welwitschia-mirabilis
No plants anchor… Hamilton & Seely, 1976, p. 284.
rain may never fall… Louw, 1972, p. 298.
current that has its origins in Antarctica… Ibid., p. 299.
five-millimetre drops that can roll… Parker & Lawrence, 2001, p. 33.
consume a third of its body weight… Hamilton & Seely, 1976, p. 285. Max: 34%.
'obsessive about their fog-basking'… Mitchell et al., 2020.
'repulsive' horned and warty skin… Bentley & Blumer, 1961.
network of straw-like capillaries… Comanns et al., 2015.
so flattened, an adaptation to eat ants… Phil Withers, University of Western Australia, speaking to In Situ Science, Ep. 27, Thorny Devils, pangolins and other outliers.
bulldozes a trench… Seely & Hamilton, 1976.
'ingestion of fog-soaked sand'… Mitchell et al., 2020.
double the amount of water collection from fog… Bhushan, 2019.
only five collect water from the fog… Mitchell et al., 2020.
arid regions account for up to 40 per cent of Earth… D'Odorico, Porporato & Runyan, 2019, p. 3.
driest place on Earth… Eshel et al., 2021, p. 1.
a territory so bleak and desolate… Morong, 1891, p. 39.
1,000 species of plants… Squeo et al., 2008.
As if aware that they have an ephemeral life… Morong, 1891, p. 40.
broad flowers the colour of lavender… Ibid., p. 42.
'a disappearance into the burrow so sudden'… Vorhies & Taylor, 1922, p. 14.
'They do not feed on cactus'… Schmidt-Nielsen & Schmidt-Nielsen, 1949, p. 181.
differ by 0.0014 millilitres… Nagy & Gruchacz, 1994, p. 1474.
'counter-current heat exchanger'… Schmidt-Neilsen, 1962, p. 24.
kangaroo rats can drink seawater… Schmidt-Nielsen & Schmidt-Nielsen, 1949, p. 183.

spend the day underground… Schmidt-Nielsen & Schmidt-Nielsen, 1950.
'explosive heat rise'… Schmidt-Neilsen, 1962, p. 16.
lose 30 per cent of their body's water… Ibid.
syphoning off the water… Ibid.
bodies to heat up to 41°C… Ibid. p. 14.
four to five litres per day can be saved… Ibid.
over 41°C by the evening… Hetem et al., 2012, p. 442, Table 2.
sun-dried leaves of *Disperma* shrubs… Taylor, 1969, p. 95.
'these plants are so dry their leaves fall apart'… Ibid.
DIDWIW trend paradigm… Feng & Zhang, 2015.
worst drought to hit the Carrizo Plain… Prugh et al., 2018.
'barren plain nearly devoid of vegetation…' Ibid.
'disturbance agent that opens niche space…' Ibid.
320 or so species… Podrabsky & Wilson, 2016, p. 500: 'the African Nothobranchiidae (over 70 species), and the South American Rivulidae (over 250 species)'.
'cloud fish'… Calviño et al., 2023, p. 1.
'exceptionally thick'… Hartmann & Englert, 2012.
turn into a form of biological glass… Podrabsky, Carpenter & Hand, 2001, R130.
survive for over 100 days… Ibid., R126.
humidity levels close to 98 per cent to survive… Fishman et al., 2008.
reach sexual maturity within 14 days… Vrtílek et al., 2018.
'Even a pool that desiccated'… Ibid., R823.
last for over a month without a whiff of oxygen… Podrabsky et al., 2007, p. 2255.
no signs of developmental abnormality… Ibid., p. 2255.

Chapter 2. Breathless (Oxygen)

blocks any sunlight from penetrating… Olson, 1932, p. 286.
breathe through their skin and a specialized anal sac… Musacchia, 1959.

'The condition of these anoxic animals'... Ultsch & Jackson, 1982, p. 23.

oxygen-based metabolism can generate ten times more power... Rich, 2003. 30 molecules of ATP compared to 2 molecules of ATP with anaerobic.

better survive prolonged periods of anoxia... Warren & Jackson, 2016.

Turtles evolved some 230 million years ago... Cleary, 2020.

From the frigid coast of Nova Scotia... Ultsch, 2006.

'turtles are several times more tolerant'... Belkin, 1963.

turtles dissolve their bony shells... Jackson, 1997.

acts as an antacid... Jackson, 2004.

See Note 3: we don't produce lactic acid... Hochachka & Mommsen, 1983.

Our own brain only weighs 2 per cent... 20 per cent of the oxygen we breathe... Pamenter, 2008, p. 2.

GABA... most common inhibitory neurotransmitter... Lari & Buck, 2021.

Over 90 per cent of neurons... Gasiorowska, 2021.

A cell, sensing the fall in oxygen, panics... Hochachka, 1986.

'reperfusion injury'... Galli & Richards, 2014.

'mitochondrial permeability transition pore'... Pamenter, 2008, p. 3.

1,200 metres... holding their breath for upwards of an hour... Tyack et al., 2006.

elephant seal... two hours... Hindell et al., 1991.

The spade-toothed whale... Thompson et al., 2012.

a few stranded individuals... Greenall, 2024.

'some of the world's most cryptic'... Tyack et al., 2006, p. 4239.

maximum dive time of three and a half hours... Quick et al., 2020.

'These extreme dive durations'... Ibid., p. 5.

18 species of penguins... Borboroglu & Boersma, 2013.

emperor penguin... 500 metres and hold its breath for half an hour... Kooyman & Ponganis, 1998, p. 19.

Elephant seals even sleep while holding their breath... Kendall-Bar et al., 2023.

'Then they transition to REM sleep'... Roth, 2023.
In 2008, Roberto Danovaro... analysed sediment samples... Danovaro et al., 2010.
DHABs are some of the driest places... Bernhard et al., 2015, p. 2.
they look like spiky squids... Neves et al., 2014, p. 5: Fig. 5.
First discovered and named in 1983... Kristensen, 1983.
over 30 known species of Loriciferan... Neves et al., 2014.
named it *Spinoloricus cinzia* after his wife... Neves et al., 2014, p. 3.
'rain of cadavers'... Danovaro et al., 2010, p. 2.
'sediments in the neighbouring'... Ibid., p. 2.
moulting as well as ovaries holding eggs... Bernhard et al., 2015.
'a paradigm shift would be necessary'... Ibid., p. 17.
'Extraordinary claims require extraordinary evidence'... Sagan, 1979.
one study struggled to find any evidence... Bernhard et al., 2015.
'no identifiable internal organs'... Ibid., p. 5.
'bodysnatchers'... WHOI, 2016.
they can digest hydrogen peroxide... Gomaa et al., 2021.
a density 20 times as high as well-aerated places... Ibid., p. 1.
'the potential metabolic capacity'... Ibid.
our oceans have lost 2 per cent of their oxygen... Schmidtko, Stramma & Visbeck, 2017.
the regions most affected by deoxygenation... Limberg et al., 2020, p. 25.
'no other environmental variable'... Rabalais et al., 2002, p. 237.
the most famous 'dead zone'... Ibid.
tributaries draining 31 of the 48 contiguous... states... https://www.epa.gov/
three times as much nitrogen... Rabalais & Turner, 2021, p. 119.
twice as much phosphorus... Ibid., p. 119.
suffocating water the size of... New Jersey... Rabalais & Turner, 2019, p. 117. Area 20,000km^2, whereas NJ is 19,050km^2 and Wales is 20,779km^2.
a path towards recovery... Gokkon, 2018.
representative in your house... Shoubridge & Hochachka, 1980.
unique ability to turn lactic acid into ethanol... Ibid.

marl-pit pond once used to irrigate farmland… Sayer et al., 2010.
'It doesn't compete well'… Holopainen et al., 1997.
'losers'… Ibid., p. 18.
'wasteful of carbon'… Shoubridge & Hochachka, 1980, p. 309.
higher levels of apoptosis… forgot how to navigate… Lefevre, 2017.
See Note 7: This geographical limit… Ultsch, 2006, p. 351.
Somalia, Ethiopia and Kenya… Jarvis & Sherman, 2002.
colonies of up to 300 individuals… Jarvis et al., 1994, p. 47.
naked mole-rats are eusocial… Jarvis, 1981.
'the animals must cooperate'… Jarvis et al., 1994, p. 51.
sparse and infrequent food sources… Jarvis & Sherman, 2002, p. 2: 'severely food restricted during dry seasons'.
'recruitment calls'… 'high-pitched contact and aggressive chirps'… Jarvis & Sherman, 2002, p. 5.
'the current data suggest'… Buffenstein et al., 2022, p. 126.
Damaraland mole-rat… (40 individuals)… Jarvis & Bennett, 1993.
she bites, nudges and squishes… Faulkes, 1990.
See Note 9: 900 pups in 11 years… Jarvis & Sherman, 2002, p. 3.
unperturbed by oxygen levels reduced to 10 per cent… Larson & Park, 2009, p. 1634.
'the mouse [tissue] did not recover at all'… Ibid.
survive complete anoxia for 18 to 20 minutes… Park et al., 2017.
nests have been found to extend for four kilometres… Buffenstein et al., 2022, p. 118.
far more resilient than a laboratory mouse… Ivy et al., 2020.
A quarter of their entire body's muscle mass… Jarvis & Sherman, 2002.
wood chip behind their incisors… Jarvis & Sherman, 2002, p. 4.
The naked mole-rats' resistance to cancer… Wlaschek et al., 2023.
a beat every minute instead of every ten… Jackson, 2013, p. 99.

Chapter 3. Fasting & Furious (Food)

cold hard granite… Jaeger, 1948, p. 45.
'I even stroked the back feathers'… Ibid.
'We made no further attempt to be quiet'… Ibid.
following the availability of food… Bent, 1940, p. 193.
'even so much as a feather'… Jaeger, 1948, p. 45.
November 1947… Jaeger, 1949.
18°C and 20°C… Ibid., p. 105.
'No movement of the chest'… Ibid., p. 106.
'unaware of [the storm's] fury'… Ibid., p. 107.
'Poor-wills are rock seeking, hibernating birds'… Ibid., p. 108.
'the sleeping one'… 'Up in the rocks'… Ibid.
he lit his campfires… Ibid., p. 109.
talented watercolourist and harpsichord enthusiast… Burtt Jr, 2015, p. 961.
comrade holding a very large net… Ibid.
a sand-covered floor, a few shelves, and a sack… Marshall, 1955.
'remained active no matter how cold the weather'… Ibid., p. 130.
the bird had lost around 20 per cent of its body weight… Ibid., p. 132.
'They awoke in a snarling and voracious condition…' Ibid., p. 134.
In 2004, a lemur was discovered… Dausmann et al., 2004.
tenrecs spend nine months of the year hibernating… Devereaux et al., 2023, p. 2.
comprising 6,000 of the total 6,400 species… Burgin et al., 2018.
killing 96 per cent of life… Erwin, 1990, p. 70.
70 per cent of all life on land… Sahney & Benton, 2008, p. 759.
First noted from the fossil record in 1841… Ibid., p. 71.
'extraordinary renewal and novelty…' Benton & Wu, 2022, p. 2.
'were all faster and nastier'… Ibid.
–3°C for several months every year… Barnes, 1989.
20,000 snakes here… Li et al., 2007.
a species of bat, two seabirds and northern white-rumped swifts… Shine et al., 2002.

Favouring the snack-sized warblers and buntings... Ibid., p. 7.
Burmese python can increase by as much as 40 times... Secor, Stein & Diamond, 1994.
our metabolism will increase by half... Ibid.
rebuilt the digestive system in just 12 hours... Ibid., p. 703.
'maintaining the intestine uselessly'... Ibid.
'exist predominantly in a fasted state'... Secor & Carey, 2016, p. 779.
milk that contains 30 per cent fat... Baker, Harington & Symes, 1963.
snakes will convert 80 per cent of their prey... Secor, Stein & Diamond, 1994, p. 701.
polar bears turn 90 per cent of seal blubber into bear fat... Best, 1984.
20 per cent of its body weight in one sitting... Derocher, 2022.
layer of blubber between 20 to 40 centimetres thick... Ford, Werth & George, 2013, p. 701.
can migrate 1,500 kilometres every year... https://iwc.int/
live for over 200 years... Mayne et al., 2019.
fasting proceeds through three stages... Battley, 2001.
A king penguin chick... fasting for four months... Rea, 1995.
Elephant seals... don't eat for up to nine weeks... Ibid.
or spend four weeks converting their fatty reserves... Ibid.
lose over a third of their body weight... Ibid.
might not eat again for another four months... Reiter, Stinson & Le Boeuf, 1978.
'our human ancestors did not consume three'... de Cabo & Mattson, 2019, p. 2544.
A cave salamander... doesn't have to eat for a decade... Bulog et al., 2000, p. 87.
increase lifespan while reducing age-related diseases... de Cabo & Mattson, 2019.
the body is forced to become more efficient... Ibid.
'refeeding syndrome'... Hearing, 2004.
the main one is a lack of phosphorus... Ibid.

not eating for five days… Ibid.

A study from Heli Routti, Sabrina Tartu… Routti et al., 2019.

started to hunt reindeer… Stempniewicz, Kulaszewicz & Aars, 2021.

hunting seals that haul themselves out… Laidre et al., 2022.

In 2007, one female known as E7… Gill Jr, 2009, p. 449.

'These non-stop flights… new extremes for vertebrate performance'… Ibid., p. 450.

recorded in flight for 237 hours… Robbins, 2022.

Yukon–Kuskokwim Delta… Gill Jr et al., 2009, p. 453: 'among the richest in the world in terms of biomass'.

body mass of some birds is over 50 per cent fat… Piersma & Gill Jr, 1998.

'obese super athletes…' Guglielmo, 2018.

fat can release eight times the amount of energy… Ibid., p. 3.

godwits pump their bodies full of molecular guides… Ibid.

some birds can drift in and out of sleep… Rattenborg et al., 2016.

5,000 kilometres back to the Arctic tundra… Gill Jr et al., 2009.

See Note 14: 30 per cent of the mudflats of the Yellow Sea… Studds et al., 2017.

See Note 14: 100 per cent of a subspecies… Ibid.

Chapter 4. Supercool Animals (Freezing)

200 million years ago… this continental kinship started to break… Clarke, Barnes & Hodgson, 2005.

30 million years ago… this terrestrial umbilical cord was severed… Ibid.

a barrier to almost all outsiders… Ibid.

three or four degrees within a few kilometres of ocean… Beers & Jayasundara, 2015, p. 1835.

The first ice sheets… 30 million years ago… Chenuil et al., 2017.

14 million years ago… a continent of thick ice… Ibid.

130 kilometres per hour for weeks on end… Eastman, 1993.

a temperature of –89°C… Rinaldi, 2006, p. 761.

satellite-based recordings… of -98°C… Scambos et al., 2018.
aboard the RRS *Discovery* in 1902… Wienecke, 2010.
'at first sight, the look of a child'… Clayton, 1776, p. 103.
'have sex with anything that moves'… Cooke, 2018.
an evolutionary pump… Clarke & Crame, 1989.
only these 16 species lack haemoglobin… Bista, 2023.
no sharks, only a few skates, one species of codfish… Eastman, 2005.
this metallic element has narcotic-like effects… Frederich, Sartoris & Pörtner, 2001.
Half of all sponges. Three-quarters of molluscs… 90 per cent of sea spider… Eastman, 2005, p. 100.
the notothenioids… 97 per cent… Ibid.
90 per cent of all the fishy biomass… Ibid., p. 97.
'the world's most distinctive marine biota'… Ibid., p. 100.
139 known species… Bista et al., 2023.
British whalers called them 'icefish'… Rudd, 1954.
***Dacodraco*…** Eastman, 1999.
they don't need to actively transport oxygen… Sidell & O'Brien, 2006.
large hearts, wide blood vessels… Ibid.
skipjack… 2 per cent of its energy budget… Ibid.
this watery blood was less viscous and easier to pump… Ibid.
'disaptation', a trait that has made these fish less adapted… Ibid.
'antifreeze glycoproteins'… ten times as effective as… ethylene glycol… DeVries, 2017.
these proteins bind to ice crystals… Meister et al., 2018.
'biochemical cunning'… DeVries, 2017.
the unit – known as OFOBS – isn't pretty… Purser et al., 2018.
a mass breeding colony of *Neopagetopsis ionah*… Purser et al., 2022.
icefish can lay 15,000 eggs, the average is 1,500… Kock & Kellerman, 1991, p. 135.
the large eggs (four to five millimetres in diameter)… Ibid., p. 138.
In the 1970s, another species… was seen guarding… Ibid., p. 139.

See Note 3: filter-feeders cover 55 per cent of the bottom... Konecki & Target, 1989.

winter temperatures that can drop to –18°C... Larson & Barnes, 2016.

muscle fibres, the spinal fluid... surrounds the heart... Storey & Storey, 1996, p. 376.

65 per cent of the frog's water has frozen... Ibid., p. 371.

thawing... begins from the inside out... Ibid., p. 377.

nine grams of sugar per 100 millilitres of blood... Storey & Storey, 1988.

human body... four grams *in total*... Wasserman, 2009.

frost triggers adrenaline... Storey & Storey, 1988.

water wouldn't freeze until it reached –38°C... Huang et al., 2021.

apple trees, olives and wild cherry, *Pseudomonas syringae*... Kennelly et al., 2007.

dust from eroded rock and desert sand, fungal spores and pollen... Woo & Yamamoto, 2020.

60 per cent of Earth's surface... Sattler et al., 2001, p. 239.

sampled from clouds thousands of metres above the ground... Morris et al., 2013, p. 95.

transported hundreds, if not thousands, of kilometres... Ibid., p. 96.

live and reproduce in the troposphere... Sattler et al., 2001.

mixed phase clouds – ice nucleators are often key... Failor et al., 2017.

'In rain, in snow-melt water'... Morris et al., 2008, p. 328.

bioprecipitation... Sands et al., 1982; Morris et al., 2014.

-3°C – a subzero squirrel... Barnes, 1989.

fend off freezing at –150°C... Sformo et al., 2009.

form of biological glass at –60°C... Carrasco et al., 2012., p. 1221.

this beetle has been called 'unfreezable'... Ibid., p. 1220.

the black bear... reduce their metabolism during hibernation... Tøien et al., 2011.

A ground squirrel... can reduce its metabolism by up to 99 per cent... Regan et al., 2022.

Every three weeks or so, its body begins to warm up… Daan, Barnes & Strijkstra, 1991.

the 'glymphatic system'… Jessen et al., 2015.

Over half of the energy reserves… Daan, Barnes & Strijkstra, 1991.

sperm, for example, were frozen with glycerol in the 1950s… Polge & Rowson, 1952.

a horse that died 800,000 years ago… Orlando et al., 2013.

three to 12 hours, a transplant will not work… Bojic et al., 2021.

10 per cent of the need for organ transplantation… Ibid.

70 per cent of all donor organs are discarded each year… Powell-Palm et al., 2021.

'Winter is coming: the future of cryopreservation'… Bojic et al., 2021.

livers can be preserved by supercooling for 44 hours… de Vries et al., 2019.

livers were preserved…*five days* of freezing… Tessier et al., 2022.

Antarctica has warmed by 3°C… bas.ac.uk/data/our-data/publication/antarctica-and-climate-change/

evidence of cooling down to 200 metres… Auger et al., 2021.

0.3°C per decade… Ibid.

bryozoans and spirorbid worms, grew faster and larger… Ashton et al., 2017.

the bryozoans and worms were slowly dying… Clark et al., 2019.

15 per cent less ice… Gilbert & Holmes, 2024.

emperor penguins are likely to go extinct by the end of this century… Lee et al., 2022.

satellite imagery shows… Fretwell, 2024.

actually gaining ice in recent decades… Blanchard-Wrigglesworth et al., 2022.

gentoo penguins were spotted breeding… Alberts, 2022.

these birds likely evolved along the coasts of Australia… Clarke et al., 2007.

Chapter 5. Highs and Lows (Pressure)

a journey of over 3,000 kilometres… Hawkes et al., 2012, p. 3.
'most formidable mountain range in the world'… Prins & Namgail, 2017, p. 40.
average height of these peaks is around 4,500 metres… Laguë, 2017, p. 2.
'At 16,000 feet, where I breathed heavily'… Swan, 1970, p. 69.
'We… have been struck with awe'… Prins & Namgail, 2017, p. 3.
bar-headed geese… fly at night or in the early morning… Parr et al., 2019.
air temperatures can be as low as −30°C… Ibid., p. 1.
'the minimum mechanical power required for flight'… Hawkes et al., 2012.
'The most frequent symptoms and signs of CMS'… Ronen et al., 2014.
10 to 20 per cent of men… develop CMS… Ibid.
they don't seem to suffer from HAPE… Parr, Wilkes & Hawkes, 2019.
birds breathe in one continuous circuit… Hazelhoff, 1951.
Their trachea is four and a half times larger… Laguë, 2017, p. 4.
'These specialisations appear to have permitted birds'… Ibid., p. 2.
A house sparrow can breathe easily… Faraci, 1991, p. 59.
grab hold of oxygen molecules more efficiently… Laguë, 2017, p. 5.
two and a half times smaller than our own… Ibid., p. 4.
twice as much capillary as a mammal's… Hawkes et al., 2017, p. 248.
mitochondria… positioned right next to the surface… Laguë, 2017, p. 5.
That record was set in 1973 by a Rüppell's vulture… Laybourne, 1974.
these birds keep flapping… Scott et al., 2015.
91 geese, the highest altitude was 7,290 metres… Hawkes et al., 2012.

geese seem unfazed by a simulated altitude of 12,000 metres… Black & Tenney, 1980.
'there must be a good explanation'… Ibid., p. 236.
200 mph winds of the jet stream… Swan, 1970.
'the summit pyramid was boiling'… Ibid.
'driven hundreds of miles out of its course'… Ibid.
'[T]he species (or its ancestor)'… Scott et al., 2015.
'No – the birds beat the mountains'… Swan, 1970.
'dwindled to one remaining species'… Anderson & Rice, 2006.
'The parallel between terrestrial and marine systems'… Ibid., p. 132.
'The surface of the ocean teems with animal life'… Ibid.
The depths of the ocean are quite as impassable… Ibid.
'According to experiment'… Ibid.
'points to the existence'… Ibid.
'It became clear back in 1954–1956'… Beliaev, 1989, p. 3.
Out of the 400-plus species of snailfish… Eschmeyer's Catalog, 2024. (https://researcharchive.calacademy.org/research/ichthyology/catalog/Species)
Mariana snailfish… (discovered and described by Gerringer… in 2014)… Gerringer et al., 2017.
the deepest fish on Earth… Lu, 2023.
'They're the deepest fish in the world'… Ibid.
hadal snailfish emerged around 20 million years ago… Wang et al., 2019.
occasional visitors to the hadal zone below 7,000 metres… Linley et al., 2016.
'I don't think there's any way'… Lu, 2023.
no fish can live permanently below 8,500 metres… Yancey et al., 2014.
TMAO holds water molecules away… Nasralla, 2020.
he began this study into TMAO in the 1990s… Yancey & Siebenaller, 1999.
mudslide off the coast of Japan in 1972… Beliaev, 1989, p. 41.
After polishing 66 snailfish ear bones… Gerringer et al., 2018.
blue microscopic fibre inside its stomach… Weston et al., 2020.

'This name speaks to the ubiquity of plastic'… Ibid., p. 169.
'ocean's ultimate trashcan'… Peng et al., 2020.
250,000 tonnes of plastic… Weston et al., 2020.

Chapter 6. Life in the Furnace (Heat)

'ants are everywhere, but only occasionally noticed'… Wilson & Hölldobler, 1990.
forage when the ground is over 60°C… Wehner, Marsh & Wehner, 1992, p. 586.
'all foragers leave'… Ibid.
thermal scavengers… Shi et al., 2015.
'It is, however, surprising that organisms'… Marsh, 1985.
***Melophorus* ants in Australia…** Christian & Morton, 1992.
***Ocymyrmex* ants in the Namib Desert…** Marsh, 1985.
'the most marvellous celerity'… Arnold, 1916, p. 194.
cool their body by another 6°C… Cerdá, 2001.
10,000 species of birds… Barrowclough et al., 2016.
beetles… 400,000 known species… Stork, 2018.
15,000 known species of ants… Schultheiss et al., 2022.
20 quadrillion… ants living on Earth… Ibid.
weigh more than all the wild mammals… Ibid.
It would weigh two-thirds of *all* the insects… Ibid.
'the little things that run the world'… Wilson, 1987.
forage in temperatures up to 54°C… Wehner, Marsh & Wehner, 1992.
a study published in *Science* in 2015… Shi et al., 2015.
reflected 41 per cent of the solar radiation… Ibid., p. 299.
they could run at 100 body lengths… Pfeffer et al., 2019.
a centimetre above can be 20°C cooler… Gehring & Wehner, 1995.
25 to 75 per cent of a foraging trip in these refuges… Wehner, Marsh & Wehner, 1992.
'path integration'… Knaden & Wehner, 2006.
'the brain of an ant is one of the most marvellous atoms'… Darwin, 1871.

critical thermal maximum for humans is 35°C… Sherwood & Huber, 2010.
regularly experience such hot and humid air by 2050… NASA, 2022.
'Outside workers, older adults, children'… Widernyski et al., 2017.
difficult to know exactly how many people die every year from the heat… Keatinge, 2003.
60,000 deaths in Europe alone… Ballester et al., 2023.
nearly half a million people died from heat stress… Zhao et al., 2021: 489,075 deaths.
'Heat… is a very brisk agitation'… Joule, 1850.
centre of the ball to around 46°C… Sugahara, Nishimura & Sakamoto, 2012.
First discovered in the 1960s, Hsps… Lindquist, 1986.
Arctic fish… produce these proteins at just 5°C… Lindquist & Craig, 1988, p. 632.
fruit fly, significant thermal stress begins at 33°C… Ibid.
Heat shock proteins… to the biological machinery… Evgen'ev, Garbuz & Zatsepina, 2014, p. 12.
same amount of Hsps at 25°C as at 50°C… Gehring & Wehner, 1995.
hypoxia, toxic metals, pesticides, UV radiation, viral infection… Evgen'ev, Garbuz & Zatsepina, 2014, p. 2.
Alzheimer's have been linked to… Hsp70… Bobkova et al., 2014.
Hsp27, can prevent the symptoms of disease altogether… Beretta & Shala, 2022.
130 million people around the world by 2050… Cummings, Reiber & Kumar, 2018.
'hyperthermophile Eden hypothesis'… Nisbet & Sleep, 2001, p. 1086.
habitable conditions began around 3.8 billion years ago… Ibid., p. 1084.
the 'Big Splat'… Ibid., p. 1083.
air temperatures rose to 2,000°C… Sleep et al., 2001.
The atmosphere was rich in carbon dioxide and low in oxygen… Ibid.

small islands of land poking through… Korenaga, 2021.
life may have emerged in the superheated, molten chaos… Sleep et al., 2001.
Dressed in brown trousers… UW-Madison, 2017.
the hottest temperature that life… can withstand is 73°C… Kemper, 1963.
'Whatever its molecular basis'… Ibid., p. 1319.
Brock published his findings… Brock & Freeze, 1969.
hottest hyperthermophile is 121°C… Takai et al., 2008.
the actual limit may be closer to 150°C… Cowan, 2004.
'PCR revolutionized everything'… McDonald, 2019.
By 420 million years ago… Bowman et al., 2009, p. 481.
oxygen in the air around 30 per cent… Wade et al., 2019.
20 metres tall… horsetails… Feng, Zierold & Röbler, 2012.
even damp wood could ignite… Bowman et al., 2009.
birds actually pick up burning sticks… John, 2018.
roughly 135 million years ago… He et al., 2012.
Oxygen levels were over 26 per cent… Ibid., p. 755.
'one of the most flammable periods'… Ibid., p. 756.
A 15 millimetre layer of bark… Ibid., p. 752.
'dark, damp, and disturbed'… Ibid., p. 756.
'There's a lot of cultural burning practice'… Robbins, 2024.
'niche construction'… Schwilk, 2003.
'Fire-spawned stands of lodgepole pine'… Vaillant, 2023, p. 198.
Schwilk investigated the dead branches of chamise… Schwilk, 2003.
What was once a 30-year-event… www.californiachaparral.org/fire
'initiating a downward ratchet'… Turner et al., 2019, p. 11325.
'Should the feedback loop of heating and drying continue'… Vaillant, 2023.
the strawberry plantations that use 85 per cent of the water… Sharrock, 2022.
number of birds in the park were at their lowest… ICTS Doñana, 2023.

Chapter 7. Ain't No Sunshine (Darkness)

ANGUS... https://www.whoi.edu/feature/history-hydrothermal-vents/discovery/1977.html
Dive number 713... Ibid.
'Isn't the deep ocean supposed to be like a desert?'... Ibid.
***Riftia pachyptila*, published in *Science*...** Jones, 1980.
'no way in, and no way out'... Yong, 2017.
every worm had over 200,000 tiny tentacles... Jones, 1980.
'It's clear!'... Yong, 2017.
'characteristic lobes of the trophosomal tissue'... Cavanaugh et al., 1981.
a 'scientific Captain Nemo'... Cromie, 1996.
the ecosystems that they furnish vary around the world... Van Dover et al., 2001.
a unique pair of light-sensitive 'organs'... Van Dover et al., 1989.
can detect the infra-red radiation of 350°C water... Pelli & Chamberlain, 1989.
'the Hoff crab'... Amos, 2015.
'very funny, very cool and endearing'... Morse, 2012.
discovered in 2006, the scaly-foot snail... Van Dover et al., 2001.
'dramatic dragon-like animal'... Chen et al., 2015.
Chong Chen, took a look inside these snails... Ibid.
an outgrowth that acted like catalytic converters... Okada et al., 2019.
In 1998, such an event happened off the coast of Oregon... Huber, Butterfield & Baross, 2003.
ROPOS... Huber, Butterfield & Baross, 2002.
200 or more different types of bacteria... Huber et al., 2006., p. 94.
photographed huge spouts of ice erupting from its southern pole... NASA, 2017.
first spotted this moon on 29 August 1789... Uri, 2024.
Tugged and pulled by the gravitational forces... Choblet et al., 2017.

ten kilometres deep… https://science.nasa.gov/mission/cassini/science/enceladus/

speed of a jet plane and reaches hundreds of kilometres into space… NASA, 2017.

Cassini had a taste of Enceladus's insides… Ibid.

bacteria 'in a [two-million-year-old] Pliocene rock… Lipman, 1928.

Throughout the middle of the twentieth century, similar findings… Kennedy, Reader & Swierczynski, 1994.

'a very tenacious grip on life.'… Ibid., p. 2526.

'it starts the first cell division'… Dombrowski, 1963, p. 460.

'stifled further research for many years'… Kennedy, Reader & Swierczynski, 1994, p. 2513.

living in carbon-rich rocks heated to 60°C… Ross, 2012.

A couple of scientists at Marie Curie's laboratory… Le Caër, 2011.

'You don't need light, food, or anything else'… Ross, 2012.

a species they named *mephisto*… Borgonie et al., 2011.

'disconnected from the entirety of traditional biology'… Colwell, Lloyd & Pratt, 2021.

a third of all the biomass on Earth… Edwards, Becker & Colwell, 2012, p. 552.

In 1986, a similar situation was discovered… Sarbu, Lascu & Brad, 2019.

There was only 7 per cent oxygen… Brad, Iepure & Sarbu, 2021, p. 1.

As of the last survey in 2020, 37 of the 57 species… Ibid.

first study on the cave… Sarbu, Kane & Kinkle, 1996.

caesium and iodine… never found in Movile Cave… Sarbu, Lascu & Brad, 2019, p. 431.

'Humans are colour-blind at night'… Kelber, Balkenius & Warrant, 2002.

100 *million* times lower than a sunlit day… Warrant, 2004, p. 766.

Kelber repeated these experiments with blue discs… Kelber, Balkenius & Warrant, 2002.

'Reliable discrimination of blue from grey is impossible'… Ibid., p. 924.

guided by a mere five photons… Interview with Eric Warrant, 23 May 2023.

Chapter 8. Taste of a Poison Paradise (Radiation)

since 1998, herds – or, more specifically, harems… Gashchak & Paskevych, 2019.
a healthy population of 150 individuals… Ibid.
'wild boar, roe deer, a red squirrel'… gorizaola.wordpress.com/blog/
traced back to 13 individuals… Xia et al., 2014.
returned to the wild from captivity in 1992… Boyd & Bandi, 2002.
over 20 tonnes of uranium-235… Kiselev, 1995.
practitioners used this as a measure of the dose… Inkret et al., 1995.
In 1898… her discovery of two new elements… Blum, 2010.
Her death 30 years later… aplastic anaemia… Ibid.
21 May 1946, Louis Slotin… Wellerstein, 2016.
'three-dimensional sunburn'… Ibid.
'Those men who allowed themselves to fear radiation'… Higginbotham, 2019.
'could spread it on bread'… Ibid.
Fifty people died… 4,000 more… https://www.iaea.org/newscenter/pressreleases/chernobyl-true-scale-accident
they're fine. They've lost that DNA damage… Clark-Hatchel et al., 2024.
they have increased by 300-fold… Ibid.
Dsup – or damage suppressor… Chavez et al., 2019.
In May 1997… Nelli Zhdanova travelled to the site… Zhdanova et al., 2000, p. 1421.
plastic hazmat suits… plastic tubing connected… Ibid., p. 1422: Figure 4.
pressing cotton buds on the walls and ceilings… Ibid., p. 1422.
37 species of fungi… Ibid., p. 1421.
'According to our calculations, the radiation doses'… Ibid., p. 1425.

'Chernobyl is, fortunately, unique'... Ibid.

80 per cent of the fungi... Ibid.

'There may be a correlation [of pigmentation]'... Ibid.

'grew towards ionizing radiation in 86 per cent of cases'... Zhdanova et al., 2004.

In 2007, Ekaterina Dadachova and Arturo Casadevall... radiosynthesis... Dadachova et al., 2007.

found that tree frogs... were darker... Burraco & Orizaola, 2022.

two and half times higher... Ibid., p. 10. Based on colony forming units (CFUs), a measure of population growth on an agar plate.

'There are recent reports that certain life'... Ibid., p. 11.

able to survive 'lethal doses of ionizing radiation'... Revskaya et al., 2012.

in 2020, her lab found that using soluble melanin... Malo et al., 2022.

One research group grew single-celled organisms... Baldwin & Grantham. 2015; Smith et al., 2015.

In 2006, a study led by Kazuo Sakai... Otsuka et al., 2006.

the 'hormesis hypothesis'... Doss, 2018.

a can of minced beef in Oregon in 1954... Anderson et al., 1956.

the cell wall of *D. radiodurans* is multi-layered... Liu et al., 2023.

forced it into a resilient microbe comparable to *D. radiodurans*... Harris et al., 2009.

repair systems are 'biologically malleable'... Byrne et al., 2014.

microbes sampled from the Sonoran Desert... Rainey et al., 2005.

'proved difficult... because they were too shy and fast'... King et al., 2015.

place radioactive waste in tropical forests... Lovelock, 2001.

Epilogue

'the greatest pollution crisis'... Margulis & Sagan, 1997, p. 108.

'When exposed to oxygen and light'... Ibid.

'The evolution of oxygen-producing cyanobacteria'... Sosa Torres, Saucedo-Vázquez & Kroneck, 2015, p. 6.

'Although trees today'... Brannen, 2017, p. 76.

80 per cent of the world... Bond et al., 2023, p. 6.

costs of the photovoltaic cells... fell by 85 per cent... IRENA, 2021.

On 5 December 2023... 192 high-powered lasers... https://lasers.llnl.gov/about/how-nif-works

likens the first fusion reaction... www.youtu.be/2kh6Ik4-yag?si=7N9aHDjaGIrYRkGX

***Wright Flyer* was airborne for 12 seconds**... Uri, 2018.

Agriculture contributes between 25 and 30 per cent... Dimbleby, 2021, p. 73.

air travel, which adds less than 4 per cent... Ibid., p. 73.

A study into 55,000 people living in the UK... Scarborough et al., 2023.

'the relationship between environmental impact'... Ibid.

'Rather than solely reacting to near-term drivers'... Anderson et al., 2023, p. 5.

'conserving nature's stage'... Ibid., p. 2.

Bibliography

Alberts, E.C. 'As climate change melts Antarctic ice, gentoo penguins venture further south'. *Mongabay*. 21 January 2022.

Amos, J. '"Hoff crab" gets formal scientific name'. *BBC News*. 25 June 2015.

Anderson, A.W. et al. (1956) Studies on a radio-resistant micrococcus. I. Isolation, morphology, cultural characteristics, and resistance to gamma radiation. *Food Technol.* 10(1): 575–577.

Anderson, M.G. et al. (2023) A resilient and connected network of sites to sustain biodiversity under a changing climate. *PNAS*. 120(7): e2204434119.

Anderson, T.R. & Rice, T. (2006) Deserts on the sea floor: Edward Forbes and his azoic hypothesis for a lifeless deep ocean. *Endeavor*. 30(4): 131–137.

Arnold, G. (1916) A Monograph of the Formicidae of South Africa. Part II. Ponerinae, Dorylinae. *Ann. S. Afr. Mus.* 14: 159–270.

Ashton, G.V. et al. (2017) Warming by 1°C drives species and assemblage level responses in Antarctica's marine shallows. *Curr. Biol.* 27: 2698–2705.

Auger, M. et al. (2021) Southern Ocean in-situ temperature trends over 25 years emerge from interannual variability. *Nat. Comm.* 12: 514.

Baker, B.E., Harington, C.R. & Symes, A.L. (2011) Polar Bear Milk I. Gross Composition and Fat Constitution. *Can. J. Zool.* 41: 1035–1039.

Baldwin, J. & Grantham, V. (2015) Hormesis: Historical and Current Perspectives. *J. Nucl. Med. Technol.* 43(4): 242–246.

Ballester, J. et al. (2023) Heat-related mortality in Europe during the summer of 2022. *Nat. Med.* 29: 1857–1866.

Barnes, B.M. (1989) Freeze Avoidance in a Mammal: Body Temperatures Below 0°C in an Arctic Hibernator. *Science.* 244(4912): 1593–1595.

Barrowclough, G.F. et al. (2016) How Many Kinds of Birds Are There and Why Does It Matter? *PLoS ONE.* 11(11): e0166307.

Battley, P.F. (2001) Is Long-Distance Bird Flight Equivalent to a High-Energy Fast? Body Composition Changes in Freely Migrating and Captive Fasting Great Knots. *Physiol. Biochem. Zool.* 74(3): 435–449.

Beers, J.M. & Jayasundara, N. (2015) Antarctic notothenioid fish: what are the future consequences of 'losses' and 'gains' acquired during long-term evolution at cold and stable temperatures? *J. Exp. Biol.* 218(12): 1834–1845.

Beliaev, G.M. (1989) *Deep-Sea Ocean Trenches and their Fauna.* Nauka Publishing House.

Belkin, D.A. (1963) Anoxia: Tolerance in Reptiles. *Science.* 139: 492–493.

Bent, A.C. (1940) Life histories of North American cuckoos, goatsuckers, hummingbirds, and their allies. Orders Psittaciformes, Cuculiformes, Trogoniformes, Coraciiformes, Caprimulgiformes, and Micropodiiformes. Bulletin of the United States National Museum. i–viii, 1–506.

Bentley, P.J. & Blumer, W.F.C. (1962) Uptake of Water by the Lizard, *Moloch horridus. Nature.* 194: 699–700.

Benton M.J. & Wu F. (2022) Triassic Revolution. *Front. Earth Sci.* 10:899541. p. 2.

Beretta, G. & Shala, A.L. (2022) Impact of Heat Shock Proteins in Neurodegeneration: Possible Therapeutical Targets. *Ann. Neurosci.* 29: 71–82.

Bernhard, J.M. et al. (2015) Metazoans of redoxcline sediments in Mediterranean deep-sea hypersaline anoxic basins. *BMC Biol.* 13: 105.

Best, R.C. (1984) Digestibility of ringed seals by the polar bear. *Can. J. Zool.* 63: 1033–1036.

Bhushan, B. (2019) Bioinspired water collection methods to supplement water supply. *Phil. Trans. R. Soc. A.* 377: 20190119.

Bista, I., et al. (2023) Genomics of cold adaptations in the Antarctic notothenioid fish radiation. *Nat. Comm.* 14: 3412.

Black, C.P. & Tenney, S.M. (1980) Oxygen transport during progressive hypoxia in high-altitude and sea-level waterfowl. *Resp. Physiol.* 39: 217–239.

Blanchard-Wrigglesworth, E. et al. 'New Perspectives on the Enigma of Expanding Antarctic Sea Ice'. *Eos.* 11 February 2022.

Blum, D. (2010) *The Poisoner's Handbook: Murder and the Birth of Forensic Medicine in Jazz Age New York.* Penguin Press.

Bobkova, N.V. et al. (2014) Therapeutic Effect of Exogenous Hsp70 in Mouse Models of Alzheimer's Disease. *J. Alz. Dis.* 38: 425–435.

Bojic, S. et al. (2021) Winter is coming: the future of cryopreservation. *BMC Biol.* 19:56.

Bond, K. et al., (2023) X-Change Electricity: On track for net zero. *RMI.* 1–37.

Borboroglu, P.G. & Boersma, P.D. (2013) *Penguins, Natural History and Conservation.* University of Washington Press.

Borgonie, G. et al. (2011) Worms from hell: Nematoda from the terrestrial deep subsurface of South Africa. *Nature.* 474(7349): 79–82.

Bowman, D. et al. (2009) Fire in the Earth System. *Science*. 324: 481–484.

Boyd, L. & Bandi, N. (2002) Reintroduction of takhi, *Equus ferus przewalskii*, to Hustai National Park, Mongolia: time budget and synchrony of activity pre- and post-release. *Appl. Anim. Behav. Sci.* 78: 87–102.

Brad, T., Iepure, S. & Sarbu, S.M. (2021) The Chemoautotrophically Based Movile Cave Groundwater Ecosystem, a Hotspot of Subterranean Biodiversity. *Diversity*. 13(128): 1–13.

Brannen, P. (2017) *The Ends of the World: Volcanic Apocalypses, Lethal Oceans, and Our Quest to Understand Earth's Past Mass Extinctions*. Oneworld Publications.

Brock, T.D. & Freeze, H. (1969) *Thermus aquaticus* gen. n. and sp. n., a Non-sporulating Extreme Thermophile. *J. Bacteriol.* 98: 289–297.

Buffenstein, R. et al. (2022) The naked truth: a comprehensive clarification and classification of current 'myths' in naked mole-rat biology. *Biol. Rev.* 97: 115–140.

Bulog, B. et al. (2000) Biology and functional morphology of *Proteus anguinus*. *Acta. Biol. Slov.* 43: 85–102.

Burgin, C.J. et al. (2018) How many species of mammals are there? *J. Mammal.* 99: 1–14.

Burraco, P. & Orizaola, G. (2022) Ionizing radiation and melanism in Chornobyl tree frogs. *Evol. Appl.* 15(9): 1469–1479.

Burtt Jr, E.H. (2015) Joe T. Marshall, Jr., 1918–2015. *Auk*. 132(4): 961–962.

Byrne, R.T. et al. (2014) Evolution of extreme resistance to ionizing radiation via genetic adaptation of DNA repair. *eLife*. 3: e01322.

Calabrese, E. & Baldwin, L. (2003) Toxicology rethinks its central belief. *Nature*. 421: 691–692.

Calviño, P. et al. (2023) A new species of Killifish from temporary aquatic environments in Northeast Argentina that stands out for its turquoise iridescence: *Argolebias guarani*. *Documentos de Divulgación*. 14: 1–6.

Carr, A. 'Journey to an End of the World'. *The New York Times*. 5 February 1961.

Carrasco, M.A. et al. (2012) Investigating the deep supercooling ability of an Alaskan beetle, *Cucujus clavipes puniceus*, via high throughput proteomics. *J. Proteomics*. 75: 1220–1234.

Carson, R. (1962) *Silent Spring*. Houghton Mifflin.

Cavanaugh, C.M. et al. (1981) Prokaryotic cells in the hydrothermal vent tube worm *Riftia pachyptila* Jones: Possible chemoautotrophic symbionts. *Science*. 213: 340–342.

Cerdá, X. (2001) Behavioural and physiological traits to thermal stress tolerance in two Spanish desert ants. *Etología*. 9: 15–27.

Chavez, C. et al. (2019) The tardigrade damage suppressor protein binds to nucleosomes and protects DNA from hydroxyl radicals. *eLife*. e47682: 1–20.

Chen, C. et al. (2015) The heart of a dragon: 3D anatomical reconstruction of the 'scaly-foot gastropod' (Mollusca: Gastropoda: Neomphalina) reveals its extraordinary circulatory system. *Front. Zool.* 12: 13, 1–16.

Chenuil, A. et al. (2017) Understanding processes at the origin of species flocks with a focus on the marine Antarctic fauna. *Biol. Rev.* 93: 481–504.

Choblet, G. et al. (2017) Powering prolonged hydrothermal activity inside Enceladus. *Nat. Astron.* 1: 841–847.

Christian, K.A. & Morton, S.R. (1992) Extreme Thermophilia in a Central Australian Ant. *Physiol. Zool.* 65(5): 885–905.

Clark, M.S. et al. (2019) Lack of long-term acclimation in Antarctic encrusting species suggests vulnerability to warming. *Nat. Comm.* 10: 3383.

Clark-Hatchel, C.M. et al. (2024) The tardigrade *Hypsibius exemplaris* dramatically upregulates DNA repair pathway genes in response to ionizing radiation. *Curr. Biol.* 34: 1819–1830.

Clarke, A. Barnes, K.A. & Hodgson, D.A. (2005) How isolated is Antarctica? *TREE*. 20: 1–3.

Clarke, A. & Crame, J.A. (1989) The origin of the Southern Ocean marine fauna. In *Origins and Evolution of the Antarctic Biota*, vol. 47, J. A. Crame (ed.). Special Publications of Geological Society of London, 253–268.

Clarke, J.A. et al. (2007) Paleogene equatorial penguins challenge the proposed relationship between biogeography, diversity, and Cenozoic climate change. *PNAS.* 104: 11545–11550.

Clayton, W. (1776) An Account of the Falkland Islands. *Phil. Trans. Royal. Soc. Lond.* 66: 99–108.

Cleary, T.J. et al. (2020) Tracing the patterns of non-marine turtle richness from the Triassic to the Palaeogene: from origin to global spread. *Palaeontology*. 63(5): 753–774.

Colwell, R., Lloyd, K.G. & Pratt, L. 'Memorial for Tullis C. Onstott, January 12, 1955 – October 19, 2021'. Astrobiology at NASA. 15 December 2021.

Comanns, P. et al. (2015) Directional, passive liquid transport: the Texas horned lizard as a model for a biomimetic 'liquid diode'. *J. R. Soc. Interface*. 12: 20150415.

Connelly, D.P. et al. (2012) Hydrothermal vent fields and chemosynthetic biota on the world's deepest seafloor spreading centre. *Nat. Comms.* 3: 630.

Cooke, L. '"Evil" penguins: The reason you shouldn't anthropomorphize animals'. Big Think, *youtube.com*. 28 October 2018.

Cooper-Driver, G.A. (1994) *Welwitschia mirabilis* – A Dream Come True. *Arnoldia*. 54: 2–10.

Cornette, R. & Kikawada, T. (2011) the induction of anhydrobiosis in the sleeping chironomid: current status of our knowledge. *Life*. 63(6): 419–429.

Cowan, D. (2004) The upper temperature for life – where do we draw the line? *Trends. Microbiol.* 12: 58–60

Cromie, W. 'Microbiologist-Aquanaut Colleen Cavanaugh Receives Tenure'. *Harvard University Gazette*. 14 November 1996.

Crowe J.H. (2015) Anhydrobiosis: An Unsolved Problem with Applications in Human Welfare. *Subcell. Biochem.* 71: 263–280.

Cummings, J., Reiber, C. & Kumar, P. (2018) The price of progress: Funding and financing Alzheimer's disease drug development. *Alz. Dem. N.Y.* 4: 330–343.

D'Odorico, P., Porporato, A. & Runyan, C. (2019) Ecohydrology of Arid and Semiarid Ecosystems: An Introduction. In D'Odorico, P., Porporato, A. & Runyan, C. (eds), *Dryland Ecohydrology*. Springer.

Daan, S., Barnes, B.M. & Strijkstra, A.M. (1991) Warming up for sleep? – Ground squirrels sleep during arousals from hibernation. *Neurosci. Lett.* 128(2): 265–268.

Dadachova E. et al. (2007) Ionizing Radiation Changes the Electronic Properties of Melanin and Enhances the Growth of Melanized Fungi. *PLoS ONE*. 2(5): e457.

Dahl, E. (1959) Amphipoda from depths exceeding 6000 metres. *Galathea Rep*. 1: 211–242.

Dai, A. (2011) Drought under global warming. *WIREs Clim Change*. 2: 45–65.

Danovaro, R. et al. (2010) The first metazoa living in permanently anoxic conditions. *BMC. Biol.* 8: 30.

Darwin, C. (1871) *The Descent of Man, and Selection in Relation to Sex*. John Murray.

Dastych, H. (2011) *Bergtrollus dzimbowski* gen. n., sp. n., a remarkable new tardigrade genus and species from the nival zone of the Lyngen Alps, Norway (Tardigrada: Milnesiidae). *Entomol. Mitt. Zool. Mus. Hamburg*. 15(186): 335–359.

Dausmann, K.H. et al. (2004) Hibernation in a tropical primate. *Nature*. 429: 825–826.

de Cabo, R. & Mattson, M.P. (2019) Effects of Intermittent Fasting on Health, Aging and Disease. *N. Engl. J. Med.* 381: 2541–2551.

de Vries, R.J. et al. (2019) Supercooling extends preservation time of human livers. *Nat. Biotechnol.* 37(10): 1131–1136.

Derocher, A. 'Feasting Season'. *Polar Bears International*. 25 April 2022.

Devereaux, M.E.H. et al. (2023) Physiological responses to hypoxia are constrained by environmental temperature in heterothermic tenrecs. *J. Exp. Biol.* 226: jeb245324.

DeVries, A.L. 'Cold as ice: antifreeze proteins in polar fishes'. *Scientia*. 22 February 2017.

Dimbleby, H. (2021) The National Food Strategy: The Plan – July 2021.

Dombrowski, H. (1963) Bacteria from Palaeozoic Salt Deposits. *Annals. N.Y. Acad. Sci.* 453–460.

Doss, M. (2018) Are We Approaching the End of the Linear No-Threshold Era? *J. Nucl. Med.* 59(12): 1786–1793.

Eastman, J.T. (1993) *Antarctic fish Biology: Evolution in a Unique Environment*. Academic Press, Inc.

Eastman, J.T. (1999) Aspects of the biology of the icefish *Dacodraco hunteri* (Notothenioidei, Channichthyidae) in the Ross Sea, Antarctica. *Polar Biol.* 21: 194–196.

Eastman, J.T. (2005) The nature of the diversity of Antarctic fishes. *Polar Biol.* 28: 93–107.

Edwards, K.J., Becker, K. & Colwell, F. (2012) The Deep, Dark Energy Biosphere: Intraterrestrial Life on Earth. *Ann Rev. Earth. Plan. Sci.* 40: 551–568.

Eiseley, L. (1957) *The Immense Journey: An Imaginative Naturalist Explores the Mysteries of Man and Nature*. Knopf Doubleday Publishing Group.

Erwin, D.H. (1990) The End-Permian Mass Extinction. *Annu. Rev. Ecol. Syst.* 21: 69–91.

Eshel, G. et al. (2021) Plant ecological genomics at the limits of life in the Atacama Desert. *PNAS*. 118(46): e2101177118.

Evgen'ev, M.B., Garbuz, D.G. & Zatsepina, O.G. (eds) (2014) *Heat Shock Proteins and Whole Body Adaptation to Extreme Environments*. Springer.

Failor, K.C. et al. (2017) Ice nucleation active bacteria in precipitation are genetically diverse and nucleate ice by employing different mechanisms. *ISME*. 11: 2740–2753.

Faraci, F.M. (1991) Adaptations to hypoxia in birds: how to fly high. *Annu. Rev. Physiol.* 53: 59–70.

Farrant, J. & Hillhorst, H. (2022) Crops for dry environments. *Curr. Opin. Biotech.* 74: 84–91.

Faulkes, C.G. (1990) Social Suppression of Reproduction in the Naked Mole Rats, *Heterocephalus glaber*. PhD Thesis. The University of London.

Feng, H. & Zhang, M. (2015) Global land moisture trends: drier in dry and wetter in wet over land. *Sci. Rep. (Nat. Publ. Group)*. 5: 18018.

Feng, Z., Zierold, T. & Röbler, R. (2012) When horsetails became giants. *Chin. Sci. Bull.* 57: 2285–2288.

Fishman, A.P. et al. (1992) Estivation in the African Lungfish. *Proc. Am. Phil. Soc.* 136: 61–72.

Ford, T.J., Werth, A.J. & George, J.C. (2013) An Intraoral Thermoregulatory Organ in the Bowhead Whale (*Balaena mysticetus*), the Corpus Cavernosum Maxillaris. *Anat. Rec.* 296: 701–708.

Frederich, M. Sartoris, F.J. & Pörtner, H-O. (2001) Distribution patterns of decapod crustaceans in polar areas: a result of magnesium regulation? *Polar Biol.* 24: 719–723.

Fretwell, P. (2024) Four unreported emperor penguin colonies discovered by satellite. *Antarc. Sci.* 1–3. https://doi.org/10.1017/S0954102023000329.

Fujii, T. et al. (2010) A Large Aggregation of Liparids at 7703 meters and a Reappraisal of the Abundance and Diversity of Hadal Fish. *BioScience*. 60: 506–515.

Galli, G.L.J. & Richards, J.G. (2014) Mitochondria from anoxia-tolerant animals reveal common strategies to survive without oxygen. *J. Comp. Physiol. B.* 184(3): 285–302.

Gashchak, S. & Paskevych, S. (2019) Przewalski's horse (*Equus ferus przewalskii*) in the Chornobyl Exclusion Zone after 20 years of introduction. *Theriologia Ukrainica.* 18: 80–100.

Gasiorowska, A. et al. (2021) The Biology and Pathobiology of Glutamatergic, Cholinergic, and Dopaminergic Signaling in the Aging Brain. *Front. Aging. Neurosci.* 13: doi.org/10.3389/fnagi.2021.654931.

Gaunitz, C. et al. (2018) Ancient genomes revisit the ancestry of domestic and Przewalski's horses. *Science*. 360: 111–114.

Gehring, W.J. & Wehner, R. (1995) Heat shock protein synthesis and thermotolerance in Cataglyphis, an ant from the Sahara desert. *PNAS.* 92: 2994–2998.

Gerringer, M.E. et al. (2018) Life history of abyssal and hadal fishes from otolith growth zones and oxygen isotopic compositions. *Deep-Sea Res. Pt I Journal.* 132: 37–50.

Gilbert, E. & Holmes, C. (2024) 2023's Antarctic sea ice extent is the lowest on record. *Weather.* 79(2): 46–51.

Gill Jr, R.E. et al. (2009) Extreme endurance flights by landbirds crossing the Pacific Ocean/ ecological corridor rather than barrier? *Proc. R. Soc. B.* 276: 447–457.

Gokkon, B. 'Global warming, pollution supersize the oceans' oxygen-depleted dead zones'. *Mongabay.* 9 January 2018.

Goldstein, B. (2022) Tardigrades and their emergence as model organisms. *Curr. Top. Dev. Biol.* 147: 173–198.

Gomaa, F. et al. (2021) Multiple integrated metabolic strategies allow foramiferan protists to thrive in anoxic marine sediments. *Sci. Adv.* 7: eabf1586.

Greenall, R. '"World's rarest whale" washes up on NZ beach'. *BBC News.* 16 July 2024.

Greven, H. (2019) From Johann August Ephraim Goeze to Ernst Marcus: A Ramble Through the History of Early Tardigrade Research (1773 Until 1929). In Schill, R.O. (ed.), *Water Bears: The Biology of Tardigrades.* Zool. Monogr. 2. Springer Nature Switzerland.

Grinnell, J. (1917) The Niche-Relationships of the California Thrasher. *Auk.* 34(4): 427–433.

Guglielmo, C.G. (2018) Obese super athletes: fat-fueled migration in birds and bats. *J. Exp. Biol.* 221: jeb165753.

Guidetti, R. & Jönsson, K.J. (2002) Long-term anhydrobiotic survival in semi-terrestrial micrometazoans, *J. Zool. (Lond.)* 257: 181e187.

Halliday, T. (2023) *Otherlands: A World in the Making.* Penguin Books Ltd.

Hamilton, W.J. & Seely, M.K. (1976) Fog basking by the Namib desert beetle, *Onymacris unguicularis. Nature.* 262: 284–285.

Harris, D.R. et al. (2009) Directed evolution of radiation resistance in *Escherichia coli*. *J Bacteriol.* 191: 5240–5252.

Hartmann, N. & Englert, C. (2012) A microinjection protocol for the generation of transgenic killifish (Species: *Nothobranchius furzeri*). *Devel. Dyn.* 241: 1133–1141.

Hawkes, L.A. et al. (2012) The paradox of extreme high-altitude migration in bar-headed geese. *Anser andicus*. *Proc. R. Soc. B.* 280: 20122114.

Hawkes L.A. et al. (2017) Goose Migration over the Himalayas: Physiological Adaptations. In Prins H.H.T. & Namgail T. (eds), *Bird Migration Across the Himalayas: Wetland Functioning amidst Mountains and Glaciers*. Cambridge University Press, 241–253.

Hazelhoff, E.H. (1951) Structure and function of the lung of birds. *Poult. Sci.* 30: 3–10.

He, T. et al. (2012) Fire-adapted traits of Pinus arose in the fiery Cretaceous. *New Phyt.* 194: 751–759.

Hearing, S.D. (2004) Refeeding syndrome: Is underdiagnosed and undertreated, but treatable. *BMJ*. 328: 908–909.

Henschel, J.R. et al. (2018) Roots point to water sources of *Welwitschia mirabilis* in a hyperarid desert. *Ecohydrology*. e2039.

Hespeels, B. et al. (2014) Gateway to genetic exchange? DNA double-strand breaks in the bdelloid rotifer *Adineta vaga* submitted to desiccation. *J. Evol. Biol.* 27: 1334–45.

Hetem, R.S. et al. (2012) Does size matter? Comparison of body temperature and activity of free-living Arabian oryx (*Oryx leucoryx*) and the smaller Arabian sand gazelle (*Gazella subgutturosa marica*) in the Saudi desert. *J. Comp. Physiol. B.* 182: 437–449.

Higginbotham, A. (2019) *Midnight in Chernobyl: The Untold Story of the World's Greatest Nuclear Disaster*. Simon & Schuster.

Hindell, M.A., Slip, D.J. & Burton, H.R. (1991) The Diving Behavior of Adult Male and Female Southern Elephant Seals, Mirounga-Leonina (Pinnipedia, Phocidae). *Aus. J. Zool.* 39(5): 595.

Hochachka, P.W. (1986) Defense strategies against hypoxia and hypothermia. *Science*. 231: 234–241.

Hochachka, P.W. & Mommsen, T.P. (1983) Protons and anaerobiosis. *Science*. 219: 1391–1397.

Holopainen, I.J. et al. (1997) Tales of two fish: the dichotomous biology of crucian carp (*Carassius carassius* (L.)) in northern Europe. *Ann. Bot. Fenn.* 34: 1–22.

Horn, D.A., Granek, E.F. & Steele, C.L. (2019) Effects of environmentally relevant concentrations of microplastic fibers on Pacific mole crab (*Emerita analoga*) mortality and reproduction. *L&O Letters*. 5: 74–83.

Huang, S. et al. (2021) Overview of biological ice nucleating particles in the atmosphere. *Environ. Int.* 146: 106197.

Huber, J.A., Butterfield, D.A. & Baross, J.A. (2002) Temporal Changes in Archaeal Diversity and Chemistry in a Mid-Ocean Ridge Subseafloor Habitat. *Appl. Environ. Microbiol.* 68(4): 1585–1594.

Huber, J.A., Butterfield, D.A. & Baross, J.A. (2003) Bacterial diversity in a subseafloor habitat following a deep-sea volcanic eruption. *FEMS. Microbiol. Ecol.* 43(3): 393–409.

Huber, J.A. et al. (2006) Microbial life in ridge flank crustal fluids. *Environ. Microbiol.* 8: 88–99.

Huerta-Sánchez E. et al. (2014) Altitude adaptation in Tibetans caused by introgression of Denisovan-like DNA. *Nature*. 512(7513): 194–197.

ICTS Doñana (2023) Estado de la biodiversidad en Doñana. Memoria 2023. Estación Biológica de Doñana, Consejo Superior de Investigaciones Científicas. Sevilla.

Inkret, W.C. et al. (1995) A Brief History of Radiation Protection Standards. *Los Alamos Science*. (23): 116–123.

IRENA (2021) Renewable Power Generation Costs in 2020. International Renewable Energy Agency, Abu Dhabi.

Ivy, C.M. et al. (2020) The hypoxia tolerance of 8 related African mole-rat species rivals that of naked mole-rats, despite divergent ventilatory and metabolic strategies in severe hypoxia. *Acta. Physiol.* (Oxf); 228(4): e13436.

Jackson, D.C., (1997). Lactate accumulation in the shell of the turtle, *Chrysemys picta bellii*, during anoxia at 3 and 10°C. *J. Exp. Biol.* 200: 2295–2300.

Jackson, D.C. (2004) Acid-base balance during hypoxic hypometabolism. Selected vertebrate strategies. *Resp. Physiol. Neuro.* 141: 273–283.

Jackson, D.C. (2013) *Life in a Shell: A Physiologist's View of a Turtle.* Harvard University Press.

Jaeger, E.C. (1948) Does the Poor-will 'Hibernate'? *Condor.* 50(11): 45–46.

Jaeger, E.C. (1949) Further observations on the hibernation of the poor-will. *Condor.* 51(3): 105–109.

Jamieson, A.J. & Yancey, P.H. (2012) On the Validity of the Trieste Flatfish: Dispelling the Myth. *Biol. Bull.* 222: 171–175.

Jarvis, J.U.M. (1981) Eusociality in a Mammal: Cooperative Breeding in Naked Mole-Rat Colonies. *Science.* 212: 571–573.

Jarvis, J.U.M. et al. (1994) Mammalian eusociality: a family affair. *TREE.* 9: 47–51.

Jarvis, J.U.M. & Bennett, N.C. (1993) Eusociality has evolved independently in two genera of bathyergid mole-rats – but occurs in no other subterranean mammal. *Behav. Ecol. Socio.* 33: 253–260.

Jarvis, J.U.M. & Sherman, P.W. (2002) *Heterocephalus glaber. Mamm. Species.* 706: 1–9.

Jessen, N.A. et al. (2015) The Glymphatic System – A Beginner's Guide. *Neurochem. Res.* 40: 2583–2599.

John, J. 'Australian "Firehawks" use fire to catch prey'. *The Wildlife Society.* 9 February 2018.

Jones, M.L. (1980) *Riftia pachyptila* Jones: Observations on the Vestimentiferan worm from the Galapágos Rift. *Science.* 213: 333–336.

Jönsson, K.I. & Bertolani, R. (2001) Facts and fiction about long-term survival in tardigrades. *J. Zool. Lond.* 255: 121–123. p. 122.

Jönsson, K.I. et al. (2008) Tardigrades survive exposure to space in low Earth orbit. *Curr. Biol.* 18(17): R729–R731.

Jørgensen, A. & Kristensen, R.M. (2001) A New Tanarctid Arthrotardigrade with Buoyant Bodies. *Zool. Anz.* 240: 425–439.

Joule, J.M. (1850) On the mechanical equivalent of heat. *Royal. Soc. Phil. Trans.* 140: 61–82.

Keatinge, W.R. (2003) Death in heat waves. *BMJ*. 327(7414): 512–513.

Keilin, D. (1959) The Leeuwenhoek Lecture: The Problem of Anabiosis or Latent Life: History and Current Concept. *Proc. R. Soc. Lond. B.* 150: 149–191.

Kelner, A., Balkenius, A. & Warrant, E.J. (2002) Scotopic colour vision in nocturnal hawkmoths. *Nature*. 419: 922–925.

Kempner, E.S. (1963) Upper Temperature Limit of Life. *Science*. 142: 1318–1319.

Kendall-Bar, J.M. et al. (2023) Brain activity of diving seals reveals short sleep cycles at depth. *Science*. 380: 260–265.

Kennedy, M.J., Reader, S.L. & Swierczynski, L.M. (1994) Preservation records of micro-organisms: evidence of the tenacity of life. *Microbiol.* 140(10): 2513–2529.

Kennelly, M.M. et al. (2007) *Pseudomonas syringae* Diseases of Fruit Trees. *Plant Dis.* 91: 4–17. 0

King, S.R.B. et al. (2015) *Equus ferus. The IUCN Red List of Threatened Species*: e.T41763A97204950

Kiselev, A.N. (1995) How much nuclear fuel is present in the lavalike fuel-containing mass in the fourth power-generating unit of the Chernobyl nuclear power plant? *Atom. Energy.* 78: 252–255.

Knaden, M. & Wehner, R. (2006) Ant navigation: resetting the path integrator. *J. Exp. Biol.* 209: 26–31.

Kock, K-H. & Kellerman, A. (1991) Reproduction in Antarctic notothenioid fish. *Antarctic Sci.* 3(2): 125–150.

Konecki, J. & Target, T. (1989) Eggs and larvae of *Nototheniops larseni* from the Spongocoel of a Hexactinellid sponge near Hugo island, Antarctic Peninsula. *Pol. Biol.* 10: 197–198.

Korenaga, J. (2021) Was there land on the Early Earth? *Life*. 11(11): 1142.

Kristensen, R.M. (1983) Loricifera, a new phylum with Aschelminthes characters from the meiobenthos. *Z. zool. Syst. Evolut.-forsch.* 21: 163–180.

Krüger, G. et al. (2017) Opportunistic survival strategy of *Welwitschia mirabilis*: Recent anatomical and ecophysiological studies elucidating stomatal behaviour and photosynthetic potential. *Botany.* 95(12): 1109–1123.

Kukal, O., Serianni, A.S. & Duman, J.G. (1988) Glycerol metabolism in a freeze-tolerant arctic insect: an in vivo ^{13}C NMR study. *J. Comp. Physiol. B.* 158: 175–183.

Laguë, S. (2017) High Altitude champions: birds that live and migrate at altitude. *J. Appl. Physiol.* 123(4): 942–950.

Laidre, K.L. et al. (2022) Glacial ice supports a distinct and undocumented polar bear subpopulation persisting in late 21st-century sea-ice conditions. *Science.* 376: 1333–1338.

Lari, E. & Buck, L.T. 2021. Exposure to low temperature prepares the turtle brain to withstand anoxic environments during overwintering. *J. Exp. Biol.* 224: jeb242793.

Larson, D.J. & Barnes, B.M. (2016) Cryoprotectant Production in Freeze-Tolerant Wood Frogs is Augmented by Multiple Freeze-Thaw Cycles. *Physiol. Biochem. Zool.* 89(4): 340–346.

Larson, J. & Park, T.J. (2009) Extreme hypoxia tolerance of naked mole-rat brain. *NeuroReport.* 20: 1634–1637.

Laybourne, R.C. (1974) Collision between a vulture and an aircraft at an altitude of 37,000 feet. *Wilson. Bull.* 86: 461–462.

Le Caër, S. (2011) Water Radiolysis: Influence of Oxide Surfaces on H2 Production under Ionizing Radiation. *Water.* 3(1): 235–253.

Lee, J.R. et al. (2022) Threat management priorities for conserving Antarctic biodiversity. *PLoS Biol.* 20(12): e3001921.

Lefevre, S. et al. (2017) Re-oxygenation after anoxia induces brain cell death and memory loss in the anoxia-tolerant crucian carp. *J. Exp. Biol.* 220: 3883–3895.

LePrince, O. & Buitink, J. (2015) Introduction to desiccation biology: from old borders to new frontiers. *Planta.* 242: 369–378. p. 372. (Quote from Claude Bernhard, 1878.)

Li, J. et al. (2007) Influence of population distribution pattern of *Gloydius shedaoensis* Zhao on predatory rate. *J. Snake*. 19:12–16.

Limberg, K.E. et al. (2020) Ocean Deoxygenation: A Primer. *One Earth*. 2: 24–29.

Lindquist, S. (1986) The Heat-Shock Response. *Ann. Rev. Biochem.* 55: 1151–1191.

Lindquist, S. & Craig, E.A. (1988) The Heat-Shock Proteins. *Annu. Rev. Genet.* 22: 631–677.

Linley T.D. et al. (2016) Fishes of the hadal zone including new species, in situ observations and depth records of Liparidae. *Deep Sea Res. I.* 114: 99–110.

Lipman, C.B. (1928) The discovery of living micro-organisms in ancient rocks. *Science*. 68(1760): 272–273.

Little, M.P. et al. (2023) Ionising radiation and cardiovascular disease: systematic review and meta-analysis. *BMJ*. 380: e072924.

Liu, F., Li, N. & Zhang, Y. (2023) The radioresistant and survival mechanisms of *Deinococcus radiodurans*. *RADMP*. 4(2): 70–79.

Louw, G.N. (1972) The role of advective fog in the water economy of certain Namib Desert animals. *Symp. Zool. Soc. Lond.* 31: 297–314.

Lovelock, J. 'We need nuclear power, says the man who inspired the Greens'. *The Telegraph*. 16 August 2001.

Lu, D. 'Scientists find deepest fish ever recorded at 8,300 metres underwater near Japan'. *The Guardian*. 3 April 2023. Interview with Alan Jamieson.

Malo, M.E. et al. (2022) Mitigating effects of sublethal and lethal whole-body gamma irradiation in a mouse model with soluble melanin. *J. Radiol. Prot.* 17: 42.

Margulis, L. & Sagan, D. (1997) The Oxygen Holocaust. In *Microcosmos: Four Billion Years of Microbial Evolution*. University of California Press. 99–114.

Marsh, A.C. (1985) Microclimatic factors influencing foraging patterns and success of the thermophilic desert ant, *Ocymyrmex barbiger*. *Insectes Soc.* 32(3): 286–296.

Marshall Jr, J.T. (1955) Hibernation in Captive Goatsuckers. *Condor*. 57(3): 129–134.

Mathews, G.B. (1938) Tardigrada from North America. *Am. Mid. Nat.* 19(3): 619–627.

Mayne, B. et al. (2019) A genomic predictor of lifespan in vertebrates. *Sci. Rep.* 9: 17866.

McDonald, C. 'Intolerable Genius: Berkeley's Most Controversial Nobel Laureate'. *California*. 12 December 2019.

Meister, K. et al. (2018) Antifreeze Glycoproteins Bind Irreversibly to Ice. *J. Am. Chem. Soc.* 140(30): 9365–9368.

Melville, D.S., Chen, Y. & Ma, Z. (2016) Shorebirds along the Yellow Sea coast of China face an uncertain future. *Emu*. 116: 100–110.

Mitchell, D. et al. (2020) Fog and fauna of the Namib Desert: past and future. *Ecosphere*. 11(1): e02996.

Møbjerg, N. & Neves, R.C. (2021) New insights into survival strategies of tardigrades. *Comp. Bio. Phys.* 254: 110890.

Morong, T. (1891) The Flora of the Desert of Atacama. *Bull. Torrey. Bot. Club.* 18: 39–48.

Morris, C.E. et al. (2008) The life history of the plant pathogen Pseudomonas syringae is linked to the water cycle. *ISME*. 2: 321–334.

Morris, C.E. et al. (2013) The Life History of *Pseudomonas syringae. Annu. Rev. Phytopathol.* 51: 85–104.

Morris, C.E. et al. (2014) Bioprecipitation: a feedback cycle linking Earth history, ecosystem dynamics and land use through biological ice nucleators in the atmosphere. *Glob. Change. Biol.* 20: 341–351.

Morse, F. 'The Hoff, A Yeti Crab With A Very Hairy Chest, Discovered In Deep Sea Vent'. *Huffington Post UK*. 4 January 2012.

Murray, J. (1910) *British Antarctic Expedition, 1907–9. Part II. On Microscopic Life in Cape Royds*. W. Heinemann.

Musacchia, X.J. (1959) The viability of *Chrysemys picta* submerged at various temperatures. *Phys. Zool.* 32: 47–50.

Nagy, K.A. & Gruchacz, M. (1994) Seasonal Water and Energy Metabolism of the Desert-dwelling Kangaroo rat (*Dipodomys merriami*). *Physiol. Zool.* 67: 1461–1478.

NASA Science Editorial Team. 'The Moon with the Plume'. *nasa. gov.* 12 April 2017.

NASA Science Editorial Team. 'Too hot to handle: How climate change may make some places too hot to live.' 9 March 2022.

Nasralla, M. et al. (2022) A study of the interaction between TMAO and urea in water using NMR spectroscopy. *PCCP.* 24: 21216–21222.

Nelson, D., Bartels, P.J. & Guil, N. (2019) Ecology of Tardigrades. In Schill, R.O. (ed.), *Water Bears: The Biology of Tardigrades.* Zool. Monogr. 2. Springer Nature Switzerland.

Nisbet, E.G. & Sleep, N.H. (2001) The habitat and nature of early life. *Nature.* 409: 1083–1091.

Okada, S. et al. (2019) The making of natural iron sulfide nanoparticles in a hot vent snail. *PNAS.* 116(41) 20376–20381.

Oliver, M. et al. (2020) Desiccation Tolerance: Avoiding Cellular Damage During Drying and Rehydration. *Ann Rev. Plant. Ecol.* 71: 7.1–7.26.

Olson, T.A. (1932) Some Observations on the Interrelationships of Sunlight, Aquatic Plant Life and Fishes. *Trans. Am. Fish. Soc.* 62: 278–289.

Orlando, L. et al. (2013) Recalibrating Equus evolution using the genome sequence of an early Middle Pleistocene horse. *Nature.* 499: 74–78.

Otsuka, K. et al. (2006) Activation of Antioxidative Enzymes Induced by Low-Dose-Rate Whole-Body γ Irradiation: Adaptive Response in Terms of Initial DNA Damage. *Rad. Res.* 166(3): 474–478.

Pamenter, M. (2008) Mechanisms of Channel Arrest and Spike Arrest underlying metabolic depression and the remarkable anoxia tolerance of the freshwater painted turtle (*Chrysemys picta bellii*). PhD Thesis. University of Toronto.

Park, T.J. et al. (2017) Fructose-driven glycolysis supports anoxia resistance in the naked mole-rat. *Science.* 356: 307–311.

Parker A.R. & Lawrence, C.R. (2001) Water Capture by a Desert Beetle. *Nature.* 414: 33–34.

Parr, N. et al. (2019) Tackling the Tibetan plateau in a down suit: insights into thermoregulation by bar-headed geese during migration. *J. Exp. Biol.* 222: jeb203695.

Parr, N., Wilkes, M. & Hawkes, L.A. (2019) Natural Climbers: Insights from avian physiology at high altitude. *High. Alt. Med. Biol.* 20(4): 427–437.

Pelli, D.G. & Chamberlain, S.C. (1989) The visibility of 350°C black-body radiation by the shrimp *Rimicaris exoculata* and man. *Nature.* 337: 460–461.

Peng, G. et al. (2020) The ocean's ultimate trashcan: Hadal trenches as major depositories for plastic pollution. *Water Res.* 168: 115121.

Pfeffer, S.E. et al. (2019) High-speed locomotion in the Saharan silver ant, *Cataglyphis bombycina. J. Exp. Biol.* 222: jeb198705.

Piersma, T. & Gill Jr, R.E. (1998) Guts don't fly: small digestive organs in obese bar-tailed godwits. *Auk.* 115(1): 196–203.

Podrabsky, J.E. & Wilson, N.E. (2016) Hypoxia and Anoxia Tolerance in the Annual Killifish *Austrofundulus limnaeus. Integrat. Comp. Biol.* 56(4): 500–509.

Podrabsky, J.E., Carpenter, J.F. & Hand, S.C. (2001) Survival of water stress in annual fish embryos: dehydration avoidance and egg envelope amyloid fibers. *Am. J. Physiol. Regulatory. Integrative. Comp. Physiol.* 280: R123–131.

Podrabsky, J.E., et al. (2007) Extreme anoxia tolerance in embryos of the annual killifish *Austrofundulus limnaeus*: insights from a metabolomics analysis. *J. Exp. Biol.* 210: 2253–2266.

Polge, C. & Rowson, L. (1952) Fertilizing Capacity of Bull Spermatozoa after Freezing at -79° C. *Nature.* 169: 626–627.

Powell-Palm, M.J. et al. (2021) Isochoric supercooled preservation and revival of human cardiac microtissues. *Comm. Biol.* 4: 1118.

Prins, H.H.T. & Namgail, T. (eds). (2017) *Bird Migration Across the Himalayas: Wetland Functioning amidst Mountains and Glaciers*. Cambridge University Press.

Prugh, L.R. et al. (2018) Ecological winners and losers of extreme drought in California. *Nat. Clim. Change*. 8: 819–824.

Purser, A. et al. (2018) Ocean Floor Observation and Bathymetry System (OFOBS): A new Towed Camera/Sonar System for Deep-Sea Habitat Surveys. *IEEE J. Ocean. Eng.* 44: 87–99.

Purser, A. et al. (2022) A vast icefish breeding colony discovered in the Antarctic. *Curr. Biol.* 32(4): 842–850.

Quick, N.J. et al. (2020) Extreme diving in mammals: first estimates of behavioural aerobic dive limits in Cuvier's beaked whales. *J. Exp. Biol.* 225: jeb222109.

Rabalais, N.N. & Turner, R.E. (2021) Gulf of Mexico Hypoxia: Past, Present, and Future. *L&O Bulletin*. 28(4): 117–124.

Rabalais, N.N., Turner, R.E. & Wiseman Jr, W.J. (2002) Gulf of Mexico, A.K.A. 'The Dead Zone'. *Annu. Rev. Ecol.* 33: 235–263.

Rainey, F.A. et al. (2005) Extensive diversity of ionizing-radiation-resistant bacteria recovered from Sonoran Desert soil and description of nine new species of the genus *Deinococcus* obtained from a single soil sample. *Appl Environ Microbiol.* 71(9): 5225–5235.

Rattenborg, N.C. et al. (2016) Evidence that birds sleep in mid-flight. *Nat. Comm.* 7: 12468.

Rea, L.D. (1995) Prolonged Fasting in Pinnipeds. PhD Thesis. University of Alaska Fairbanks.

Rebecchi, L., Boschetti, C. & Nelson, D. (2020) Extreme-tolerance mechanisms in meiofaunal organisms: a case study with tardigrades, rotifers and nematodes. *Hydrobiologia*. 847: 2779–2799, p. 2790.

Rebecchi, L. et al. (2011) Resistance of the anhydrobiotic eutardigrade *Paramacrobiotus richtersi* to space flight (LIFE–TARSE mission on FOTON-M3). *J. Zool. Syst. Evol.* Res. 49: 98–103.

Regan, M.D. et al. (2022) Nitrogen recycling via gut symbionts increases in ground squirrels over the hibernation season. *Science*. 375: 460–463.

Reiter, J., Stinson, N.L. & Le Boeuf, B.J. (1978) Northern elephant seal development: The transition from weaning to nutritional independence. *Behav. Ecol. Sociobiol.* 3: 337–367.

Revskaya, E. et al. (2012) Compton scattering by internal shields based on melanin-containing mushrooms provides protection of gastrointestinal tract from ionizing radiation. *Cancer Biother. Radiopharm.* 27(9): 570–576.

Rich, P. R. (2003) The molecular machinery of Keilin's respiratory chain. *Biochem. Soc. Trans.* 31: 1095–1105.

Rinaldi, A.C. (2006) The cold side of life. *EMBO*. 7: 759–763.

Robbins, J. 'The Godwit's Long, Long Nonstop Journey'. *The New York Times*. 20 September 2022.

Robbins, J. 'To Protect Giant Sequoias, They Lit a Fire'. *The New York Times*. 9 July 2024.

Ronen, R. et al. (2014) The genetic basis of chronic mountain sickness. *Physiol.* 29(6): 403–412.

Ross, V. 'Discover Interview: Tullis Onstott Went 2 Miles Down & Found Microbes That Live on Radiation'. *Discover Magazine*. 26 June 2012.

Roth, A. 'Elephant Seals Take Power Naps During Deep Ocean Dives'. *The New York Times*. 20 April 2023.

Routti, H. et al. (2019) State of knowledge on current exposure, fate and potential health effects of contaminants in polar bears from the circumpolar Arctic. *Sci. Total. Environ.* 664: 1063–1083.

Rudd, J.T. (1954) Vertebrates without Erythrocytes and Blood Pigment. *Nature*. 173: 848–850.

Sagan, C. (1979) *Broca's Brain: Reflections of the Romance of Science*. Random House Publishing Group.

Sahney, S. & Benton, M.J. (2008) Recovery from the most profound mass extinction of all time. *Proc. R. Soc. B.* 275: 759–765.

Sands, D.C. et al. (1982) The association between bacteria and rain and possible resultant meteorological implications. *J. Hung. Met. Serv.* 86: 148–152.

Sarbu, S.M., Kane, T.C. & Kinkle, B.K. (1996) A Chemoautotrophically Based Cave Ecosystem. *Science*. 272: 1953–1955.

Sarbu, S.M., Lascu, C. & Brad, T. (2019) Dobrogea: Movile Cave. In Ponta, G.M.L. & Onac, B.P. (eds), *Cave and Karst Systems of Romania*. Springer International Publishing AG.

Sarkissian, C. et al. (2015) Evolutionary genomics and conservation of the endangered Przewalski's horse. *Curr. Biol.* 25(19): 2577–83.

Sattler, B. (2001) Bacterial growth in supercooled cloud droplets. *Geophys. Res. Lett.* 28(2): 239–242.

Sayer, C.D. et al. (2010) The decline of crucian carp *Carassius carassius* in its native English range: the example of rural ponds in north Norfolk. ECRC Research Report.139: 1–19.

Scambos, T.A. et al. (2018) Ultralow Surface Temperatures in East Antarctica from Satellite Thermal Infrared Mapping: The Coldest Places on Earth. *Geophys. Res. Lett.* 45(12): 6124–6133.

Scarborough, P. et al. (2023) Vegans, vegetarians, fish-eaters and meat-eaters in the UK show discrepant environmental impacts. *Nat. Food.* 4: 565–574.

Schmidt-Neilsen, B. & Schmidt-Nielsen, K. (1949) The Water Economy of Desert Mammals. *The Scientific Monthly.* 69(3): 180–185.

Schmidt-Nielsen, B. & Schmidt-Nielsen, K. (1950) Evaporative water loss in desert rodents in their natural habitat. *Ecology.* 31: 75–85.

Schmidt-Neilsen, K. (1962) Comparative Physiology of Desert Mammals. *Agricultural Experiment Station.* Special Report 21.

Schmidtko, S., Stramma, L. & Visbeck, M. (2017) Decline in global oceanic oxygen content during the past five decades. *Nature*. 542: 335–339.

Schultheiss, P. et al. (2022) The abundance, biomass, and distribution of ants on Earth. *PNAS*. 119(4): e2201550119.

Schwilk, D.W. (2003) Flammability Is a Niche Construction Trait: Canopy Architecture Affects Fire Intensity. *Am. Nat.* 162: 725–733.

Scott, G.R. et al. (2015) How Bar-Headed Geese Fly Over the Himalayas. *Physiol.* 30(2): 107–115.

Secor, S.M. & Carey, H.V. (2016) Integrative Physiology of Fasting. *Compr. Physiol.* 6(2): 773–825.

Secor, S.M., Stein, E.D. & Diamond, J. (1994) Rapid upregulation of snake intestine in response to feeding: a new model of intestinal adaptation. *Am. J. Physiol.* 266: G695–705.

Seely, M.K. & Hamilton, W.J. (1976) Fog catchment sand trenches by Tenebrionid beetles, *Lepidochora*, from the Namib Desert. *Science.* 193: 484–486.

Sformo, T. et al. (2009) Deep supercooling, vitrification and limited survival to −100°C in the Alaskan beetle *Cucujus clavipes puniceus* (Coleoptera: Cucujidae) larvae. *J. Exp. Biol.* 213: 502–509.

Sharrock, D. 'Strawberry farmers can take their pick of illegal wells at Spain's Doñana reserve'. *The Times*. 14 January 2022.

Sherwood, S.C. & Huber, M. (2010) An adaptability limit to climate change due to heat stress. *PNAS.* 107(21): 9552–9555.

Shi, N.N. et al. (2015) Keeping cool: Enhanced optical reflection and radiative heat dissipation in Saharan silver ants. *Science.* 349(6245): 298–301.

Shine, R. et al. (2002) A review of 30 years of ecological research on the Shedao Pitviper, *Gloydius shedaoensis. Herpetol. Nat. Hist.* 9: 1–14.

Shoubridge, E.A. & Hochachka, P.W. (1980) Ethanol: Novel End Product of Vertebrate Anaerobic Metabolism. *Science.* 209(4453): 308–309.

Sidell, B.D. & O'Brien, K.M. (2006) When bad things happen to good fish: the loss of hemoglobin and myoglobin expression in Antarctic icefishes. *J. Exp. Biol.* 209(10): 1791–1802.

Slack, H.J. (1851) *Marvels of Pond Life: or a year's microscopic recreations among the polyps, infusoria, rotifers, water-bears, and polyzoa*. Groombridge & Sons.

Sleep, N.H., Zahnle, K. & Neuhoff, P.S. (2001) Initiation of clement surface conditions on the earliest Earth. *PNAS*. 98: 3666–3672.

Sloan, D., Batista, R.A. & Loeb, A. (2017) The Resilience of Life to Astrophysical Events. *Sci. Rep.* 7: 5419.

Smith, G.F. et al. (2011) Exploring biological effects of low level radiation from the other side of background. *Health Phys.* 100(3): 263–265.

Solnit, R. (2005) *Hope in the Dark: Untold Histories, Wild Possibilities*. Canongate Books.

Somero, G.N. & DeVries, A.L. (1967) Temperature Tolerance of Some Antarctic Fishes. *Science*. 156: 257–258.

Sosa Torres, M.E., Saucedo-Vázquez, J.P. & Kroneck, P.M.H. (2015) The Magic of Dioxygen. In Kroneck, P. & Sosa Torres, M. (eds), *Sustaining Life on Planet Earth: Metalloenzymes Mastering Dioxygen and Other Chewy Gases*. Metal Ions in Life Sciences, vol 15. Springer, Cham.

Squeo, F.A., Arancio, G. & Gutiérrez, J.R. (eds) (2008) *Libro Rojo de la Flora Nativa y de los Sitios Prioritarios para su Conservación: Región de Atacama*. Ediciones Universidad de La Serena, 6: 97–120.

Stempniewicz, L., Kulaszewicz, I. & Aars, J. (2021) Yes, they can: polar bears *Ursus maritimus* successfully hunt Svalbard reindeer *Rangifer tarandus platyrhynchus*. *Polar Biol.* 44: 2199–2206.

Storey, K.B. & Storey, J.M. (1988) Freeze tolerance in Animals. *Physiol. Rev.* 68: 27–84.

Storey, K.B. & Storey, J.M. (1996) Natural Freezing Survival in Animals. *Annu. Rev. Ecol. Syst.* 27: 356–386.

Stork, N.E. (2018) How Many Species of Insects and Other Terrestrial Arthropods Are There on Earth? *Ann. Rev. Ent.* 63(1): 31–45.

Studds, C. et al. (2017) Rapid population decline in migratory shorebirds relying on Yellow Sea tidal mudflats as stopover sites. *Nat. Commun.* 8: 14895.

Sugahara, M. Nishimura, Y. & Sakamoto, F. (2012) Differences in Heat Sensitivity between Japanese Honeybees and Hornets under High Carbon Dioxide and Humidity Conditions inside Bee Balls. *Zool. Sci.* 29: 30–36.

Swan, L.W. (1970) Goose of the Himalayas. *Nat. Hist.* 79: 68–75.

Takai, K. et al. (2008) Cell proliferation at 122 degrees C and isotopically heavy CH4 production by a hyperthermophilic methanogen under high-pressure cultivation. *PNAS.* 105(31): 10949–10954.

Taylor, C.R. (1969) The Eland and the Oryx. *Scientific American.* 220(10): 89–95.

Taylor, W.T.T & Barrón-Ortiz, C.I. (2021). Rethinking the evidence for early horse domestication at Botai. *Sci. Rep.* 11: 7440.

Tessier, S.N. et al. (2022) Partial freezing of rat livers extends preservation time by 5-fold. *Nat. Comm.* 13: 40008.

Thompson, K. et al. (2012) The world's rarest whale. *Curr. Biol.* 21: Pr905–R906.

Tøien, Ø. et al. (2011) Hibernation in black bears: independence of metabolic suppression from body temperature. *Science.* 331(6019): 906–909.

Tsujimoto, M., Imura, S. & Kanda, H. (2015) Recovery and reproduction of an Antarctic tardigrade retrieved from a moss sample frozen for over 30 years. *Cryobiology.* 72: 78–81.

Turner, M.G. et al. (2019). Short-interval severe fire erodes the resilience of subalpine lodgepole pine forests. *PNAS.* 116(23): 11319–11328.

Tyack, P.L. et al. (2006) Extreme diving of beaked whales. *J. Exp. Biol.* 209: 4238–4253.

Ultsch, G.R. (2006) The ecology of overwintering among turtles: where turtles overwinter and its consequences. *Biol. Rev.* 81: 339–367.

Ultsch, G.R. & Jackson, D.C. (1982) Long-Term Submergence at 3°C of the Turtle, *Chrysemys Picta Bellii*, in Normoxia and Severely Hypoxic Water. I. Survival, Gas Exchange and Acid-Base Status. *J. Exp. Biol.* 96: 11–28.

Uri, J. '115 years ago: Wright brothers make history at Kitty Hawk'. *nasa.gov.* 17 December 2018.

Uri, J. '235 Years Ago: Herschel Discovers Saturn's Moon Enceladus'. *nasa.gov*. 29 August 2024.

UW-Madison. 'Rediscovering Yellowstone – Tom Brock'. *youtube.com*. 28 April 2017.

Vaillant, J. (2023) *Fire Weather: A True Story from a Hotter World*. Knopf Doubeday Publishing Group.

Van Dover, C.L. et al. (1989) A novel eye in 'eyeless' shrimp from hydrothermal vents of the Mid-Atlantic ridge. *Nature*. 337: 458–460.

Van Dover, C.L. et al. (2001) Biogeography and Ecological Setting of Indian Ocean Hydrothermal Vents. *Science*. 294: 818.

Vecchi, M. et al. (2021) The toughest animals of the Earth versus global warming: Effects of long-term experimental warming on tardigrade community structure of a temperate deciduous forest. *Ecol. Evol.* 11: 9856–9863.

Vorhies, C.T. & Taylor, W.P. (1922) Life History of the Kangaroo Rat, *Dipodomys spectablis spectablis* Merriam. *USDA*. Bulletin No. 1019: 1–40. p. 14.

Vrtílek, M et al. (2018) Extremely rapid maturation of a wild African annual fish. *Curr. Biol.* 28: R803–R825.

Wade, D.C. et al. (2019) Simulating the climate response to atmospheric oxygen variability in the Phanerozoic: a focus on the Holocene, Cretaceous and Permian. *Clim. Past.* 15: 1463–1483.

Wang, K. et al. (2019) Morphology and genome of a snailfish from the Mariana Trench provide insights into deep-sea adaptation. *Nat. Ecol. Evol.* 3: 823–833.

Warrant, E. (2004) Vision in the dimmest habitats on Earth. *J. Comp. Physiol. A.* 190: 765–789.

Warren, D.E. & Jackson, D.C. (2016) The metabolic consequences of repeated anoxic stress in the western painted turtle, *Chrysemys picta bellii. Coimp. Biochem. Physiol. A.* 203: 1–8.

Wasserman, D.H. (2009) Four grams of glucose. *Am. J. Physiol. Endocrinol. Metab.* 296: E11–E21.

Wehner, R., Marsh, A.C. & Wehner, S. (1992) Desert ants on a thermal tightrope. *Nature*. 357: 586–587.

Wellerstein, A. 'The demon core and the strange death of Louis Slotkin'. *The New Yorker*. 21 May 2016.

Westh, P. & Ramløv. (1991) Trehalose Accumulation in the Tardigrade *Adorybiotus coronifer* During Anhydrobiosis. *J. Exp. Zool.* 258:303–311.

Weston, J.N. et al. (2020) New species of *Eurythenes* from hadal depths of the Mariana Trench, Pacific Ocean (Crustacea: Amphipoda). *Zootaxa*. 4748: 163–181.

WHOI. 'Can animals thrive without oxygen?' 28 January 2016.

WHOI. 'Discovering Hydrothermal Vents: 1977 – Astounding Discoveries'.

Widernyski, S. et al. (2017). The Use of Cooling Centers to Prevent Heat-Related Illness: Summary of Evidence and Strategies for Implementation Climate and Health Technical Report. Series Climate and Health Program. Centers for Disease Control and Prevention.

Wienecke, B. (2010) The history of the discovery of emperor penguin colonies 1902–1904. *Polar Record*. 46(238): 271–276.

Wilson, E.O. (1987) The Little Things That Run the World (The Importance and Conservation of Invertebrates). *Cons. Biol.* 1(4): 344–346.

Wilson, E.O. & Hölldobler, B. (1990) *The Ants*. Harvard University Press.

Wlaschek, M. et al. 2023. The skin of the naked mole-rat and its resilience against aging and cancer. *Mech. Ageing. Dev.* 216: 111887.

Wolff, T. (1961) The Deepest Recorded Fishes. *Nature*. 4772: 283.

Woo, C. & Yamamoto, N. (2020) Falling bacterial communities from the atmosphere. *Environ. Microbiome*. 15: 22.

Xia, C. et al. (2014). Reintroduction of Przewalski's horse (*Equus ferus przewalskii*) in Xinjiang, China: The status and experience. *Biol. Conserv.* 177: 142–147.

Yancey, P.H. & Siebenaller, J.F. (1999) Trimethylamine oxide stabilises teleost and mammalian lactate dehydrogenases against inactivation by hydrostatic pressure and trypsinolysis. *J. Exp. Biol.* 202: 3597–3603.

Yancey, P.H. et al. (2014) Marine fish may be biochemically constrained from inhabiting the deepest ocean depths. *PNAS*. 111(12): 4461–4465.

Yong, E. 'How Giant Tube Worms Survive at Hydrothermal Vents: I Contain Multitudes'. *youtube.com*. 13 November 2017.

Zhao, Qi et al. (2021) Global, regional, and national burden of mortality associated with non-optimal ambient temperatures from 2000 to 2019: a three-stage modelling study. *Lancet Planetary Health*. 5(7): e415–e425.

Zhdanova, N.N. et al. (2000) Fungi from Chernobyl: mycobiota of the inner regions of the containment structures of the damaged nuclear reactor. *Mycol. Res.* 104(12): 1421–1426.

Zhdanova, N.N. et al. (2004) Ionizing radiation attracts soil fungi. *Mycol. Res.* 108(9): 1089–1096.

Index